D0655559

Robotic Assembly

WITHDRAWN

International Trends in Manufacturing Technology

ROBOTIC ASSEMBLY

Edited by Professor Keith Rathmill

IFS (Publications) Ltd, UK

Springer-Verlag
Berlin Heidelberg New York Tokyo
1985

TS
178
.4
R62

British Library Cataloguing Publication Data

Robotic assembly.—(International trends in manufacturing technology)
    1. Assembling machines—Automatic control
    2. Robots. Industrial
    I. Rathmill, Keith   II. Series
    670.42'7   TS178.4

ISBN 0-903-608-71-5 IFS (Publications) Ltd
ISBN 3-540-15483-3 Springer-Verlag Berlin Heidelberg New York Tokyo
ISBN 0-387-15483-3 Springer-Verlag New York Heidelberg Berlin Tokyo

© 1985 **IFS (Publications) Ltd,** 35-39 High Street, Kempston,
                    Bedford MK42 7BT, UK
    and **Springer-Verlag** Berlin Heidelberg New York Tokyo

The work is protected by copyright. The rights covered by this are reserved, in
particular those of translating, reprinting, radio broadcasting, reproduction by
photo-mechanical or similar means as well as the storage and evaluation in data
processing installations even if only extracts are used. Should individual copies for
commercial purposes be made with written consent of the publishers then a remittance
shall be given to the publishers in accordance with §54, Para 2, of the copyright law. The
publishers will provide information on the amount of this remittance.

Phototypeset by Wagstaffs Typeshuttle, Henlow, Bedfordshire
Printed and bound by Short Run Press Ltd, Exeter

International Trends in Manufacturing Technology

The advent of microprocessor controls and robotics is rapidly changing the face of manufacturing throughout the world. Large and small companies alike are adopting these new methods to improve the efficiency of their operations. Researchers are constantly probing to provide even more advanced technologies suitable for application to manufacturing. In response to these advances IFS (Publications) Ltd is publishing a series of books on topics that highlight the developments taking place in manufacturing technology. The series aims to be informative and educational.

Subjects to be covered in the series include:

Robot vision
Programmable assembly
Robot safety
Robotic assembly
Robot sensors
Electronics assembly
Flexible manufacturing systems
Robotic welding
Automated guided vehicles

The series is intended for manufacturing managers, production engineers and those working on research into advance manufacturing methods. Each book will be published in hard cover and will be edited by a specialist in the particular fields.

This, the third in the series – Robotic Assembly – is under the editorship of Professor Keith Rathmill of the Cranfield Robotics and Automation Group. The series editors are: Michael Innes, John Mortimer, Brian Rooks, Jack Hollingum and Anna Kochan.

Finally, I express my gratitude to the authors whose works appear in this publication.

John Mortimer,
Managing Director,
IFS (Publications) Ltd

Acknowledgements

IFS (Publications) Ltd wishes to express its acknowledgement and appreciation to the following publishers/organisations for granting permission to use some of the papers reprinted within this book.

British Robot Association
28-30 High Street
Kempston
Bedford MK42 7BT
England

CIRP
International Institution for
  Production Engineering Research
19, rue Blanche
75009 Paris
France

Fraunhofer Institut für Produktions-
technik und Automatisierung (IPA)
Postfach 800 469
Nobelstrasse 12
D-7000 Stuttgart 80
West Germany

Springer-Verlag GmbH
Otto-Suhr Allee 26/28
D-1000 Berlin 10
West Germany

Society of Manufacturing
Engineers
One SME Drive
P.O. Box 930
Dearbon, MI 48128
USA

IFS (Conferences) Ltd
35-39 High Street
Kempston
Bedford MK42 7BT
England

PERL Hitachi Ltd
292 Yoshida-Machi
Totsuka-ku
Yokohama 244
Japan

The Institution of Mechanical
Engineers
1 Birdcage Walk
Westminster
London SW1
England

Contents

1. Assembly Robots

2. Robot Assembly Systems and Applications

3. Product Design for Robot Assembly

4. Programming Systems

5. Sensory Systems

6. Economics

Preface

The industrial robot really began by making inroads in spot welding applications over two decades ago. Since then it has found further growth points notably in the fields of arc welding and machine servicing. As these and other uses of the industrial robot are subjected to ongoing development, it is clear that the wide and potentially far-reaching field of assembly is now establishing itself as the prime future growth area for the industrial application of robots over the next decade.

The distinction between robotic assembly as we currently perceive it and conventional assembly automation is perhaps best seen from the standpoint that established high-speed, special-purpose, high-volume, single-product-orientated assembly machines are now being complemented by highly versatile, easily reprogrammable, multi-product, low-volume orientated, off-the-shelf assembly robots. The assembly robot can therefore be usefully visualised as an essential component of future flexible assembly systems which will be capable of producing families of similar products in a dynamic and responsive fashion.

Frequently the requirements of an industrial assembly task include speed and precision. As the steadily increasing variety of industrial robot variants evolve, adapt and are further developed to perform satisfactorily against these needs, there will undoubtedly be many further developments in robot hardware and software technology.

One of the foremost important perspectives of this subject is an understanding of just how early on in the developing field of robotic assembly we actually are. This book is the result of an earnest attempt to provide amongst other things an accurate statement of the 'state-of-the-art'. In so doing hopefully it will help to arouse in the reader a keen awareness of the short and dynamic history supporting today's developments. It is sobering to reflect that little more than ten years have passed since the introduction of the world's first commercially available electrically powered robot, the ASEA IRb6. It was not until 1979 that I had my first hands-on experience of the Unimation PUMA. More recently we have seen the parallel development of a number of Cartesian assembly robots, and with a clearly gathering momentum towards a specialisation in assembly, the popularisation of the numerous SCARA style machines. As a final comment on this ongoing evolutionary trend we cannot fail to note the highly significant introduction, late in 1984, of the ASEA 1000 pendular assembly robot (discussed in Chapter 1). Those who saw it unveiled at the SCANAUTOMATIC '84 Exhibition in Sweden would, I am sure, recognise it as yet another milestone along a continuing road of assembly robot evolution.

So the subject of assembly robotics is far from having a plateau, and current developments in product design, applications expertise, gripper design, drive systems, control systems, feeder design, sensing and not least robot programming systems would seem to offer ever increasing support for the view that we have not yet begun to 'scratch the surface' of the future potential in robotic assembly.

Amidst all this evolutionary dynamics there are, of course, good reasons which currently motivate companies to consider applying available robot systems technology to their assembly requirements. Amongst these reasons must be counted the very real benefits of greater management control and improved quality consistency in batch assembly operations. There are a number of such commercially tangible reasons which are identified, for example by Michael Leete (Chapter 2) in his Flymo Case Study. However, not least important is the more general argument that 'now and not later' is the time to begin in gaining in-house experience of this strategically important technology. Staff training is not considered within the scope of this book but it is nonetheless one of the major benefits to be derived from timely and well managed applications of robotic assembly. I maintain the view that the greatest single retarding influence in the development of any company's manufacturing expertise and efficiency is not connected directly with the sophistication of the technology. It is rather more the challenge created by the 'rate of change' of that technology and the consequent requirement that is placed upon the company to constantly update its manufacturing expertise base.

The text does set out to strike a sensible balance between the 'state-of-the-art' and relevant examples of forward looking research and development of an industrially relevant nature. In Chapter 1 an attempt has been made to provide complementary views of assembly robots from a balanced selection of those available. I do hope that the numerous reputable robot suppliers and individual experts whose views are not included in one or more of the following chapters will understand the challenge that Mike Innes and I faced in striving to provide the reader with a balanced perspective in preference to attempting a 'who's who' of assembly robotics.

The hard truth is that very often the most telling component of any new technological expertise is that of applications know-how. I have therefore a special interest in the reader enjoying and benefitting from the blend of case studies we have 'assembled' in Chapter 2.

Given the very high profile of interest that product design for automated manufacture has been receiving from major companies in recent years, space has been committed to three different views of this important area of consideration. Whilst product design for robotic assembly is still a difficult area in which to provide a brief and generalised text, product design for robot automated assembly really deserves consideration and emphasis should be placed upon this activity as an integral part of a well managed robotic assembly applications project.

One of the most recurrent areas of practical challenge in the industrial experience of robotics generally is that of programming. In Chapter 4 this subject is given a deliberately forward tilt and relevant examples of 'state-of-the-art' developments are combined with overviews which hopefully illustrate both the potential of future developments and the characteristics of current diversity.

The final chapters deal with examples of relevant sensory systems developments and economic considerations, respectively. A major proportion of current international robotics research effort is being put into the development of what might be called a truly intelligent robot. Whatever limits such intellectual capability may be seen to aspire to, that ultimate sophistication and (perhaps most importantly) all the necessary if relatively humble stages of commercially relevant sophistication along the way will fundamentally require adequate sensory feedback. Few people doubt the future value of ragged, high-resolution, fast-response, low-cost, vision sensing; and examples of progress in this direction are presented together with useful force sensing developments in Chapter 5.

Chapter 6 takes a consciously analytical approach in considering economic issues and in so doing is intended to provide a useful contrast to prior sections of the book which collectively make more general remarks on the total return-on-investment issues.

In these introductory remarks I have tried to stress the dynamic and continuing evolution which characterises assembly robotics. Currently efforts are being put into better mechanical design and direct drives in order to achieve improved speed and accuracy in assembly robots. Although ongoing increases in the performance of robots are clearly expected and are surely welcomed by all, one of the most exciting prospects must be the distinct possibility of a demand for assembly robots which permits true volume production. Currently the purchase price of many assembly robots reflects the expensive early stages of market development with heavy sales support costs and low sales volumes against which to write off initial development budgets. The prospect of volume-produced, low-cost assembly robots is not without very serious foundation. This distinct possibility of significantly reducing costs combined with a steadily increasing level of operational capability and sophistication identifies the assembly robot as a technology with a lot more than a bright future.

Keith Rathmill
April 1985

1
Assembly Robots

The development of the assembly robot to date can be seen as an evolutionary phenomonon with an increasing variety of devices, some quite specialised, now available to the purchaser. This chapter examines examples of current technology together with some more recent developments in assembly robot manipulators.

PRAGMA A3000 – A SOLUTION TO THE REQUIREMENTS OF FLEXIBLE AUTOMATION

A. Camera
DEA SpA, Italy
and
S. Salmeri
DEA GmbH, West Germany

First presented in German at the MHI Congress, April 1983, Hanover. Reproduced by permission of the authors and IPA Stuttgart.

Assembly operations in many industrial areas represent one of the most important starting points for increasing production. However, measures taken for the introduction of automation have not always been crowned with success, as they have produced many problems and difficulties. The DEA PRAGMA A3000 industrial assembly robot is discussed with a view to flexible integrated assembly.

Owing to their fascinating problematic nature, assembly robots have for years attracted the attention of research technologists, but it is only recently that they have been introduced into practical production. This was done through the initiative of the few technologically advanced organisations which approached the problem realistically in order to find applications for the robot which could offer sensible immediate solutions both on the technical and the economic level. The most attractive possibilities for the use of assembly robots is in the assembly of medium quantities of small parts (also from component families) in cases where assembly operations have a higher added value, and especially where the products have not matured completely and may still be modified in the course of time.

Admittedly, the assembly robot is still in its infancy but it is getting constantly clearer that the advantages of its use are not only theoretical but also realistic and attainable. It must, of course, be kept in mind that at today's state of the technology the robot cannot take over every type of difficult assembly operation even though it is capable of overcoming numerous problems in the course of production. It should also be noted at this point that the major part of the technical difficulties are caused by problems connected with parts feeding and orientation. In addition, the cost/earnings ratio is more favourable with the introduction of robots when account is taken of the requirements of automatic assembly during the planning of the product and when the organisation of the product system retains the parts in order and correctly orientated.

Flexible integrated and 'made-to-measure' systems

The introduction of assembly robots requires in particular a global view of the problem to be solved. This is due to the fact that robots represent the most qualified element of a complicated production system to which also belong the necessary preconditions for the orientation and feeding of parts, the technical assembly processes and dimensional and functional tests.

This represents an automatic cell in which the robot generally carries out the following tasks and employing the most favourable operating sequence in order to obtain the required productivity: gripping, placing and fitting parts together. The use of several robots integrated in a single assembly cell makes increased productivity possible and indirectly offers considerable advantages. These include the possibility of integrating various secondary operations in a single homogeneous complex. In addition several cells can be interconnected to produce more elaborate assembly lines.

The philosophy of the DEA PRAGMA A3000 is based on the following basic concepts for realising an integrated assembly system by optimum means:

- Mechanical and electronic modules matched to the complexity of the application.
- Full programmability in order to guarantee flexibility and the optimisation of the assembly cycle even if the product or the product mix is altered after installation.
- Adaptability of the robot in order to activate reject or repeat cycles when abnormal conditions occur, automatically overcoming breakdowns or alerting the operators after a certain number of attempts.

The application philosophy for the PRAGMA A3000

The decision made in the planning phase of the PRAGMA A3000 was not to produce an assembled machine with a rigid structure but simply to make available to the planner of assembly systems some basic components (primarily a control unit and arms), which can easily be combined with conveyors, feeding devices, and assembly and test apparatus. By means of the interface boards housed in the control cabinet (up to 512 input and output signals), it is possible to control and monitor not only the robots but also all the peripherals.

Design solutions

The basic modules are first of all arms with linear movements along the x, y, and z axes (Fig. 1) and the control system (Fig. 2). With additional rotating axes on the wrist combined in different ways, up to five degrees of freedom are obtainable for each arm. The use of a Cartesian structure is the most

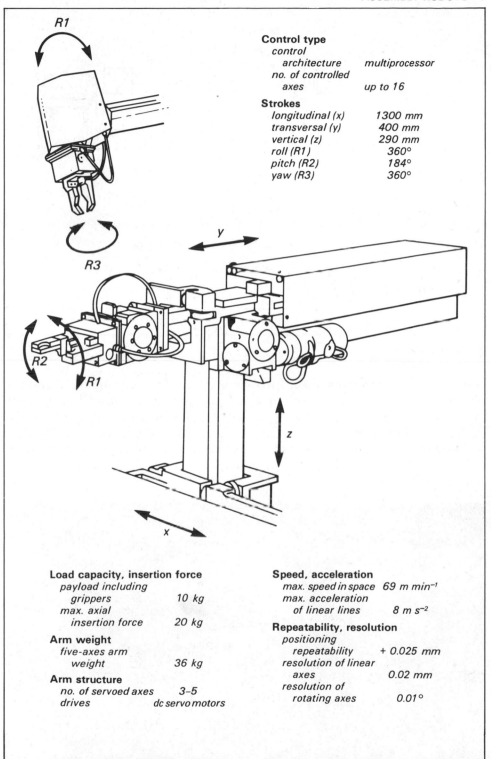

Control type
  control
    architecture   multiprocessor
  no. of controlled
    axes   up to 16

Strokes
  longitudinal (x)   1300 mm
  transversal (y)   400 mm
  vertical (z)   290 mm
  roll (R1)   360°
  pitch (R2)   184°
  yaw (R3)   360°

Load capacity, insertion force
  payload including
    grippers   10 kg
  max. axial
    insertion force   20 kg

Arm weight
  five-axes arm
    weight   36 kg

Arm structure
  no. of servoed axes   3–5
  drives   dc servo motors

Speed, acceleration
  max. speed in space   69 m min⁻¹
  max. acceleration
    of linear lines   8 m s⁻²

Repeatability, resolution
  positioning
    repeatability   + 0.025 mm
  resolution of linear
    axes   0.02 mm
  resolution of
    rotating axes   0.01°

Fig. 1   The PRAGMA A300 industrial assembly robot and technical specifications

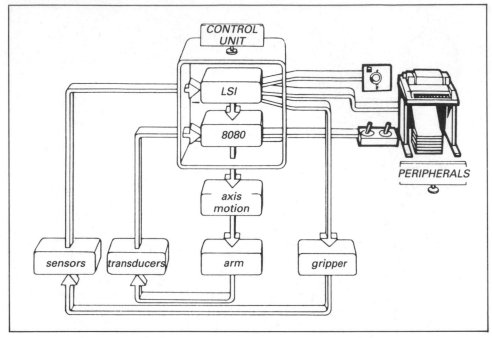

Fig. 2 Control system

suitable for solving assembly problems since for very many assembly operations the picking-up, orientation, positioning and insertion of parts takes place along preferred axes at right angles. With the horizontal and vertical version the most varied assembly systems can be arranged optimally, and horizontal and vertical arms can operate simultaneously in a cell without interfering with each other – thus producing shorter cycle times and a greater output (Figs. 3 and 4).

Fig. 3 PRAGMA A3000 two-arm configuration

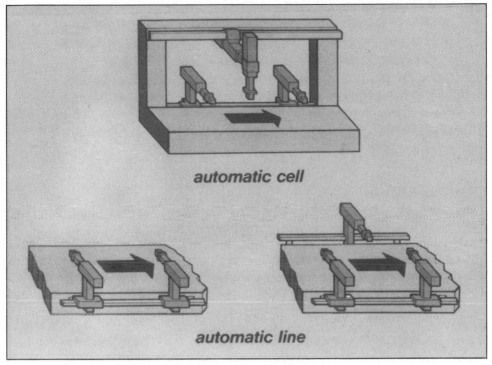

Fig. 4   Assembly cell with horizontal and portal robots

Design philosophy is in principle the same for both versions. The axes are driven by dc motors and positioned by incremental pulse generators (optical encoders) which are mounted on the motor axis. The position control is point-to-point with controlled deceleration without overshoot or oscillation. On the x axis the arm moves along a guide rail, available in different lengths, which can be fitted to any support.

The control unit housed in the control cabinet can control up to four arms and monitor and control sensors, grippers, peripherals and assembly devices through input/output modules. It is a multiprocessor unit with different functions at two levels. The upper level is concerned with the execution of the user program and is based on the micorcomputer LSI 11/23, while the lower level (using an 8080 microprocessor) carries out the point-to-point positioning control of the axes. The information exchange between the two microcomputers goes through a common memory.

Programming

The programs for the PRAGMA A3000 can be divided as follows:

- RT-11 computer operating system.
- Robot operating system.
- User program.

All these programs are present in the mass storage and before execution are loaded into the main memory.

RT-11 computer operating system

The RT-11 computer operating system makes the system programs available. In particular it supports the writing, editing, modifying and storing of the programs.

Robot operating system

The robot operating system supports the execution of the user programs which are written in the language HELP. The operating system permits the execution of virtually parallel processes such as the arm movements and the driving of peripheral equipment in the assembly system.

User programs

The user programs describe the operations of the assembly system. They are written in HELP 11 (high-level expandable language for programming) which was developed by DEA for calculations and process control. It is an interactive general purpose computer language which also offers commands referring to assembly in addition to the logical commands such as IF . . . THEN-ELSE, WHILE-FOR, GOTO, GOSUB. The available commands permit the general operation of the system (HOLD-ZERO), the control of the arms (MOVE, SMOVE, SPEED, HALT) and of the peripherals (SET, RESET), and the monitoring of the sensors (IVALUE, FORCE). Owing to the completeness of this language it is also possible to write subprograms for the statistical evaluation of the production.

The user programs consist of the assembly operating program and the teaching program. The assembly operating program defines the assembly cycle. It consists of commands which determine the arm movements, the calculations and the test, as well as the activation of the inputs and outputs for the work-holding devices. A point-to-point control is used for the movements of the linear and the rotational axes. The movements can be monitored by force and/or presence sensors.

The data required for the assembly operation program are generally determined by a special teaching program during the programming phase. With these, certain coordinate values (e.g. at loading points) can be determined as well as sensor values with the aid of the robot as a measuring machine. For measuring the coordinates during the programming phase the respective positions are approached using a joystick and then stored according to the so-called 'teach-in' process.

Application areas and problems

Assembly robots can be applied to a wide range of industrial areas, in particular the automobile industry with its supply branches, and the electromechanical and precision products industries. The configuration of the system depends largely on the requirements of the user and is mostly determined by the goal to achieve the required output and return on investment.

The present technical level permits the realisation of effective solutions in which, from the economic point of view, speed, accuracy, decision capability and the easy redesign and reprogramming of the machine play a large part.

There are, of course, some restrictions on the achievable flexibility, as it is necessary to provide special mechanical devices for parts feeding and orientation. To overcome these restrictions various measures must be taken:

- Parts planning for assembly.
- Standardisation for peripherals.
- Structural organisation to maintain the parts in the correct position throughout the production process.
- Development of sensors (tactile and visual) which can contribute to solving recognition and orientation problems.

Four different assembly systems implemented with the PRAGMA robot are described briefly below. They illustrate examples of applications in the machining, automobile and electrical industries.

- *Universal joint.* The joint consists of 16 component parts (Fig. 5) which can all be gripped by the versatile gripper. The part is assembled by two arms in 21 seconds.

- *Selective assembly of compressor crankshaft.* This set-up incorporates two arms each with three degrees of freedom and one arm with two degrees of freedom. This is a selective assembly (Fig. 6) during which the fit between the crankshaft and the external bearing, arranged in five classes, is based on the measuring process by a robot at a measuring station which is integrated into the assembly system. The washer and circlip are also assembled in about 10 seconds making production of about 350 units per hour possible.

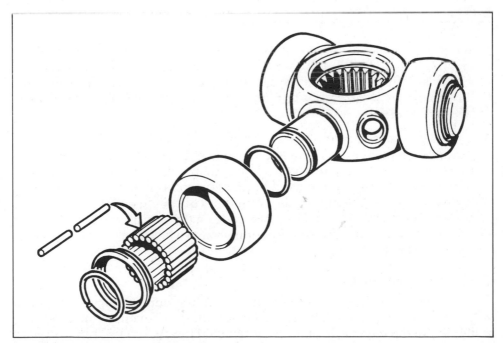

Fig. 5   Universal joint

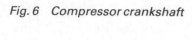

Fig. 6   Compressor crankshaft

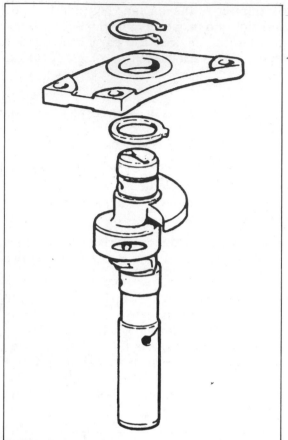

- *Assembly of an electric motor.* In this example three robots assemble an electric motor (Fig. 7) for a washing machine consisting of 20 parts in 20 seconds. The motor is also checked for performance, and checked electrically. The assembly line must assemble 10 different motor types without resetting.

- *Cylinder head assembly.* About 150 cylinder head units (Fig. 8) must be produced in one hour. The complete assembly line consists of robot cells interconnected by a pallet conveyor system. The number of arms used depends on the requirements of production and reliability. The assembly line is capable of assembling different types of engine owing to a magnetic code on the pallet. Through the first code on the transport pallet the robot selects, for example, the corresponding valve which is to be assembled to the cylinder head.

Concluding remarks

The complexity and importance of the assembly systems implemented so far with the PRAGMA A3000 are proof for the effectiveness and the high reliability of the technical solutions obtained. However, a few remarks must still be made.

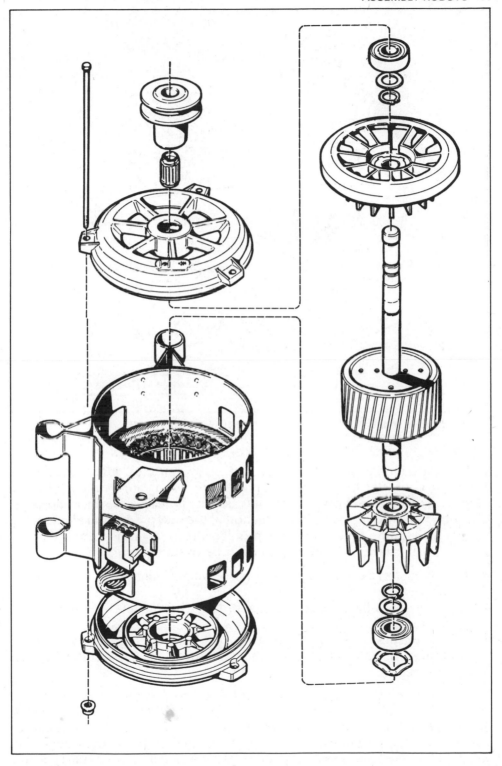

Fig. 7 Electric motor

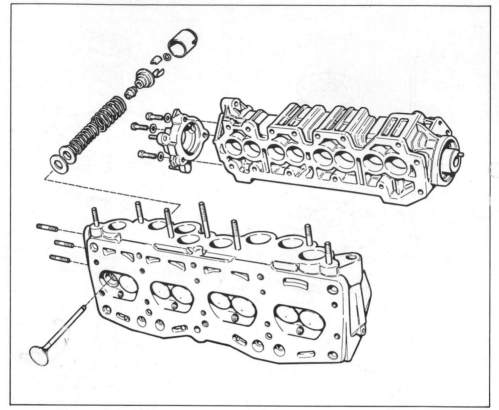

Fig. 8   Cylinder head

The introduction of robots into a factory must be made within the framework of flexible automation and not as a single event. Before employing robots an exact analysis of feasibility is necessary which shows the advantages thrown up by the introduction of the new production element but which at the same time shows the problems (e.g. regarding quality of the parts) which can occur. The variables to be examined are:

- The characteristics of the product (size, weight, number of components, variants).
- Average life of the product (degree of maturity and modifications over a period of time).
- Characteristics of the production process (existing organisation, stores, environment, state of training of the operators, working conditions, space requirements).

Finally, it should be pointed out that the improvements which stand to be gained from this investment can be considerable if the application philosophy of the robot and its characteristics are really flexible and adapted to the many application possibilities. Only then will the high degree of availability and the possibility of regaining a substantial part of the investment contribute to effective economic operation.

THE SCARA ROBOT AND ITS FAMILY

H. Makino and N. Furuya
Yamanashi University, Japan

First presented at the 3rd International Conference on Assembly Automation, 25-27 May 1982, Stuttgart, West Germany. Reproduced by permission of the authors and IFS (Conferences) Ltd.

The SCARA robot is a university-born assembly robot which has four degrees of freedom. The name stands for Selective Compliance Assembly Robot Arm and it can compensate for positional error by selective compliance. By the simple and stiff mechanism and by the 'virtual cam curve control method', it can move at a very high speed of up to 1.5m/s without any vibration. Because of the good cost – performance ratio of the robot, several companies now produce and supply some industrial versions of it. Features of the SCARA robot and its family are discussed and the typical applications are outlined.

The SCARA assembly robot has a relatively simple construction and works with high performance and flexibility[1-3]. The robot was developed in the laboratory of the Precision Engineering Department of Yamanashi University, headed by Professor Makino, and supported by a research consortium known as the SCARA Research Group. After three years development the robot could reach a high level of performance, and several companies of the Group decided to sell their own versions of the robot.

The commercial supply of the so-called 'SCARA type' robot started in April 1981. Twelve months later about 400 were installed in production lines, which was equal to the total number of existing assembly robots already installed.

Historical background

Japan now has over twenty years history of automatic assembly. Initially much was learned from the USA before the Japanese developed many kinds of elements and systems by themselves. Nowadays Japanese production engineering and automation technology are regarded as among the best in the world. However, these developments are limited to the field of fixed automation. The industries dealing with mass production advanced the most in automation, whereas middle or small-sized companies retained manual assembly. Even in large companies, the product mix has become larger because of customer demand and technical innovation. There is a trend to more variety and small lot sizes, and consequently machines with more flexibility are needed.

The Automatic Assembly Committee of the Japan Society of Precision Engineering (JSPE), conducted by the Chairman, Professor Taniguchi of Tokyo Science University, has been discussing the problem of flexible assembly for over ten years. Some people claimed that the building block system (BBS) was the answer to flexibility; others argued for the intelligent robot. However, Professor Makino thought that the practical answer was somewhat between the two. It might even be numerical control – the prototype of the NC assembly centre was built in 1974[4]. Professor Heginbotham (then of Nottingham University, UK) seemed to point in the same direction[5]. Although the speed of the NC prototype was too slow, the possibility of such a system was encouraging.

Experiments in insertion practice using force sensors and feedback control were then conducted. Hitachi had already developed their own method [6], while Nevins and Whitney were studying the mating phenomena[7] and later RCC (remote centre compliance)[8].

Through the experiments, a problem became clear: in some cases and in some directions stiffness is needed, whereas in others compliance is needed. This problem led to the idea of *selective compliance*.

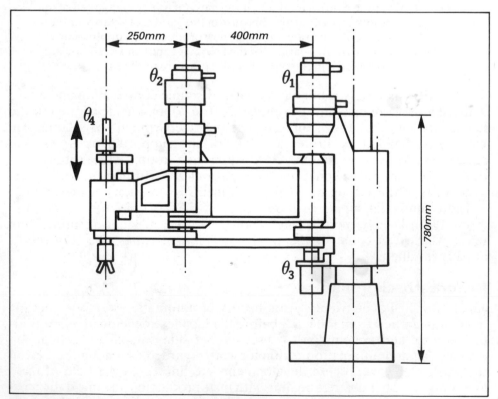

Fig. 1   Construction and dimensions of SCARA

Principle of selective compliance

The basic construction of the SCARA is shown in Fig. 1. The dimensions given are of the second prototype but the construction of the first and second is the same. The SCARA has four degrees of freedom:

- Horizontal rotation of the joints θ_1 and θ_2 deciding planar position of the tool point.
- Rotation θ_3 deciding the orientation of the tool about the z axis.
- The last degree of freedom θ_4 for the vertical motion of the tool.

This kind of construction is suitable for the assembly task of 'down-to-up' mounting, which is considered to form 80% of existing work if the product design is suitable.

The two arms, or links, drived by the two dc servo motors make a 'Byobu'-like structure. Byobu, a folded parting plate, is traditional in Japanese furniture made of wood and paper. This structure is very rigid under vertical loading. Incorporating this design gives the machine the feature of selective compliance, i.e. the selectively different compliance by direction. The SCARA has small compliance in the z direction and x and y moments, and large compliance in the x and y directions and the z moment.

The effect of the selective compliance is shown in Fig. 2. If there is a chamfer on the peg or on the hole, and the peg touches the chamfer portion, then the peg will move laterally by the effect of lateral compliance. However, if there is a tilting compliance as well, the peg may tilt and jamming will occur when the clearance is small. With selective compliance, lateral compliance is large and tilting compliance small, and positional correction without jamming is possible. This ensures good performance in assembly.

Through laboratory experiments[1], and through the practical experience by the commercial model, it has become evident that the SCARA can compensate for positional error less than the chamfer dimension, that is, so long as the tool or component strikes the chamfered surface on insertion.

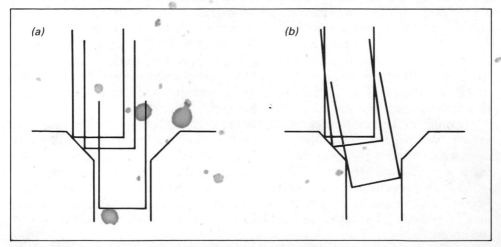

Fig. 2   Effect of selective compliance: (a) lateral compliance and (b) tilting compliance

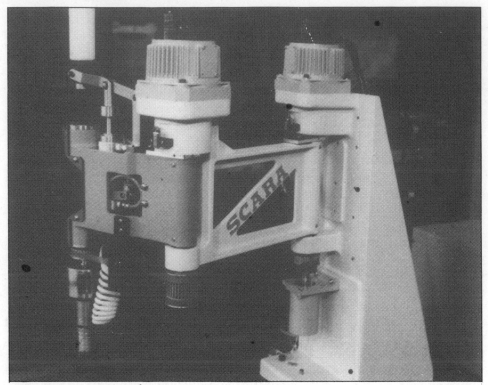

Fig. 3 First SCARA prototype

Development of the SCARA

The first prototpye SCARA was designed in January 1978. As previously mentioned the SCARA Research Group was set up to finance the work. Members of the Group were entitled to obtain any information on the hardware and software of the SCARA and were given the right to freely use the two applied patents in barter with the finance. The Group started in April 1978 with five companies, and finished in March 1981 with thirteen companies. Meetings were held every two months, where results of the experiments were shown and the direction of development was discussed.

The first prototype model, shown in Fig. 3, was produced in August 1978. The major characteristics of the selective compliance were checked on the machine. The machine had a problem in dynamic characteristics and later the motors and servo-drivers were changed.

Despite the rather poor motion characteristics, the effectiveness of the system was evident. It was thought that the robot might suit many assembly operations, especially screw driving. Thus it was decided to make the second prototype machine, shown in Fig. 4, mainly for the purpose of screw driving.

With the development of the second prototype in May 1980, the robot began to move at the expected high speed, i.e. 1m/s in maximum velocity and 1g (9.8m/s^2) in maximum acceleration. No vibration or overshoot was observed.

Since then, experimental verification of dynamic characteristics was made

Fig. 4   Second SCARA prototype

and some practical application problems were checked. In March 1981 it was regarded that the robot had reached a practical level of development and the consortium was closed. Consequently, some Group members announced that they would sell their commercial versions.

Motion control techniques

The SCARA is controlled by an 8-bit (Z80) microcomputer. The motion of the two major axes are controlled as if they were moved by cam mechanisms. A predetermined acceleration–deceleration characteristic cam curve is memorised in the microcomputer as a digital table and recalled when moving. This method is called 'virtual cam curve control'[3]. In applying this method, the robot can move at a very high speed with a human-like smoothness.

The curve used now is the 'NC2' curve, shown in Fig. 5, which is the second curve designed for numerical control. The duration of deceleration is twice that of acceleration so vibration is prevented. The calculation of the digital data is made by a large computer using the 'universal cam curve' program written in FORTRAN, which was previously developed by one of the authors.

Features of the SCARA robot

The SCARA robot has many features to match assembly tasks. These are discussed below.

Selective compliance

By the previously mentioned selective compliance feature, the SCARA can adapt to small variations in position. It overcomes the difficulty of quality control of the parts, long set-up time of the machine, and unnecessary high resolution of positioning. Thus it assures high performance of assembly tasks.

As a compliance apparatus, RCC is well-known. The differences between the RCC and SCARA method are:

- RCC can correct both angular and lateral error, whereas SCARA can correct laterally only. However, in an existing practical application, the initial angular error seems to be sufficiently small. The problem is if angular error is caused by the insertion force.
- RCC is a wrist compliance system and its design is specified by the weight and dimensions of the gripped work. The compliance of SCARA is from the arm configuration and has no relation to the work or tool.
- Even if the tool head is changed, the compliance of SCARA is maintained.
- SCARA holds the vertical attitude of the tool. This is useful for press-fit operations.

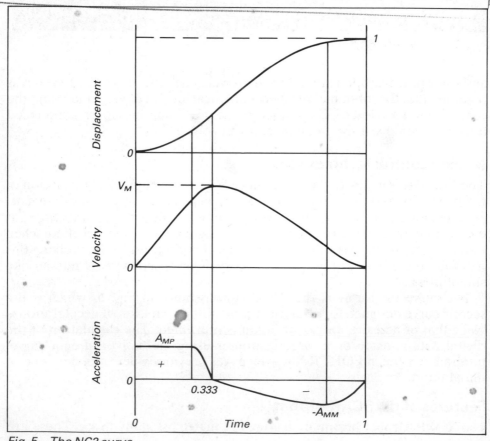

Fig. 5   The NC2 curve

Motion control

By means of virtual cam curve control, a very smooth movement is obtained with high speed. The motion is human-like in its smoothness, and more machine-like in its speed and accuracy.

Payload

Due to its rigidity in the z-direction, SCARA can bear a large load. The thrust force of the pneumatic cylinder used in the prototype is about 300N under normal pressure. This means that the tool can hold a weight up to 30kg. SCARA can thus do more work than simple handling; for example, it can be used for pressing, staking, screw-driving, nut-running, tapping, etc.

Working area

A jointed type robot can cover a larger working area than one with Cartesian coordinates. According to the authors SCARA can cover about ten-times a larger working area than a Cartesian coordinate robot of the same size. SCARA requires little space for installation and is suitable for an assembly line with some kind of conveyor.

Computer control

SCARA is controlled by a microcomputer and thus the task program is easy to change. Small lot and mixed production is possible, as is sensory feedback, if required.

Industrial versions and applications

Although the SCARA is still in its infant years, it now plays an active part in production lines. Several companies in the SCARA Research Group developed their own commercial models after the closure of the Group in March 1981. The names of these robots are different, but the configurations are very similar.

Picmat-SCARA

Picmat-SCARA (Figs. 6–8) is supplied by Nitto Seiko Co. Ltd, one of the largest screw manufacturers in Japan who also supply screw-driving machines and general purpose assembly machines. It was known throughout the experiments that the screw-driving process was suitable for this robot – the second prototype SCARA was designed by an engineer from Nitto Seiko in a week working alongside Professor Makino. Later, the design was changed slightly and a series of three sizes was made. Figs. 6 and 7 show the largest model of the Picmat-SCARA used for screw tightening of door sashes. The screw feeder is attached together with the tool to the arm. Other applications include light bulb assembly (Fig. 8).

In the first year Nitto Seiko sold over 100 robots. The major application is screw driving of screws 3–5mm in diameter with an automatic screw feeder. Typical processing times for this kind of operation is 3s; 0.5s for moving a single stroke (of 150mm, for instance), 1s for up and down motion, and 1.5s for screw driving (tool rotation).

Fig. 6   Picmat-SCARA used for door sash screw-tightening

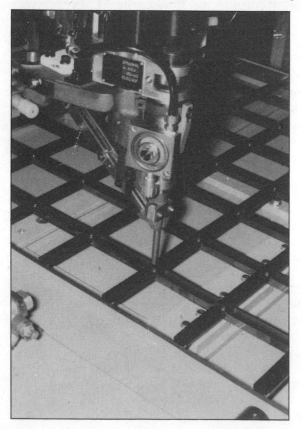

Fig. 7   Door sash screw-tightening

Fig. 8   Light bulb assembly

SKILAM

SKILAM (Figs 9–11) is possibly the best selling assembly robot. Sankyo Seiki Seisakusho Co. Ltd produces two sizes of the robot and the total number of installations in the first year was about 200. In 1982, it was producing about 30 to 50 robots per month, and the announced cooperation with IBM in the USA will double the production.

The SKILAM robot is the only one that utilises the virtual cam curve control, and thus it has the best motion characteristics. The maximum speed of the tool point reaches 1.5m/s with sufficient smoothness. It is questionable why other manufacturers do not use this method but they are investigating its adoption.

The major application of the SKILAM is pick-and-place motion and insertion. The pick-up finger is usually designed specifically for the handled parts. Fig. 10 shows the ATC (automatic tool change) tools, and Fig. 11 shows palletising or packing of music boxes.

The first and the largest user of the SKILAM is Pioneer Co. Ltd, one of the largest video and audio apparatus manufacturers. They use eight SKILAM (SCARA) robots in series in an assembly line of printed circuit boards. It is known that there are some kinds of high-speed insertion machines for proper shaped axial and radial parts but there is no flexible

Fig. 9   SKILAM in test

insertion machine for square-shaped or large-sized parts. Pioneer engineers tackled this problem; they bought standard SKILAM robots without tooling and tooled them by themselves. This, together with the reliability of the robot and its selective compliance, has given the production line an admirable working efficiency of 99.5%.

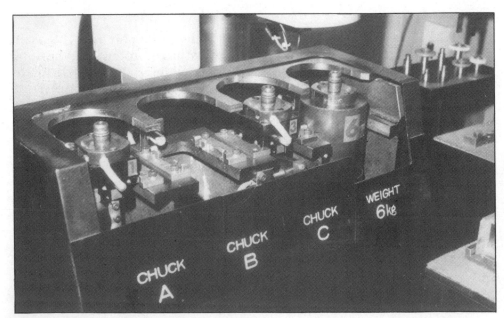

Fig. 10   ATC tools for SKILAM

Fig. 11   SKILAM packing music boxes

PUHA

Pentel Co. Ltd is a well-known manufacturer of writing goods. Initially the company developed a robot for its own use. The application is the insertion of a very small lead retaining rubber to the metal tip of a mechanical pencil (see Figs. 12 and 13). The rubber is a small (about 2mm in diameter) conical

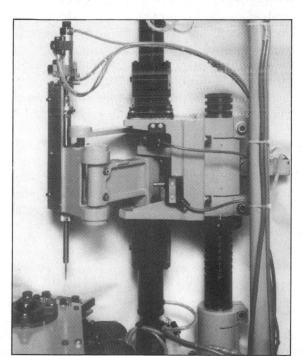

Fig. 12   PUHA used for the insertion of very small rubbers in mechanical pencils

Fig. 13   Pentel's metal tip and rubber assembly machine

part and very difficult to orientate and feed in a vibratory bowl feeder. It is moulded in a sheet in an 11 × 11 matrix. So it was decided to pick off the rubber directly from the sheet at the assembly station. Two methods were investigated: one was to use an NC controlled X-Y table, and the other the use of a robot. As there were many advantages in using a robot, a Pentel engineer together with Professor Makino designed the PUHA (Pentel Universal Handling Arm).

The robot has parallelogram links so an extra motor on the pillar replaces the elbow motor. Thus, inertia is reduced, but results in a smaller working area.

The tip assembly machine, shown in Fig. 13, consists of three components: the index table which finishes and inspects the metal tip, PUHA robot as a station unit of the machine, and an indexing conveyor to feed rubber sheets. The cycle time of this machine is 2s. The machine was built in September 1980, and after two months trial in the machine shop, it was introduced to the manufacturing division where it operates 24 hours a day.

NEC Models B and C

NEC, one of the largest computer and communications companies, has three or four types of robots. Among them the Model B and Model C are of the SCARA type. The company produces robots for in-house use as well as for sale. Fig. 14 shows an application, where two Model B robots cooperate and assemble five parts of a telephone receiver.

Fig. 14   Two NEC robots cooperating in the assembly of telephone receivers

Fig. 15   Small oil pump assembly with CAME robots

CAME

Yamaha Motor Cycle Co. Ltd also makes an assembly robot known as CAME. The robot was designed by the Yamaha engineers but the principle is the same as SCARA. As of 1982 the company had three years experience in introducing robots in the production line, and had built about 80 robots in three years. The major application of the robot is nut and bolt running in 6 – 8mm diameters. However, many kinds of applications, such as bonding, tapping, heavy press fit, and air blowing are being attempted. Fig. 15 shows one of the applications where two CAME robots are working in an assembly line for small oil pumps.

Arm-Base

Arm-Base robots are supplied by Hirata Machinery Co. Ltd, not a former member of the SCARA Research Group but who produce the robots under licence. The robot controllers are standardised for several types, so that the user can select a particular type for a specific application.

References

[1]   Makino, H. and Furuya, N. 1980. Selective compliance assembly robot arm. In, *Proc. 1st Int. Conf. on Assembly Automation*, 25-27 March 1980, Brighton, UK, pp.77-86. IFS (Publications) Ltd, Bedford, UK.

[2]   Makino, H. and Furuya, N. 1980. Research and development of the SCARA robot. In, *Proc. 4th Int. Conf. on Production Engineering*, August 1980, Tokyo, pp.885-890. Japan Society of Precision Engineering, Tokyo.

[3]   Makino, H. and Furuya, N. 1981. Motion control of a jointed arm robot utilizing a microcomputer. In, *Proc. 11th Int. Symp. on Industrial Robots*, 7–9 October 1981, Tokyo, pp.405-412. Japan Industrial Robot Association, Tokyo.

[4]   Makino, H. 1974. An experimental approach to NC assembly centre. In, *Proc. 1st Int. Conf. on Production Engineering*, August 1974, Tokyo, pp.486-491. Japan Society of Precision Engineering, Tokyo.

[5]   Heginbotham, W.B. et al. 1976. A versatile variable mission assembly machine. In, *Proc. 6th Int. Symp. on Industrial Robots*, 24–26 March 1976, Nottingham, UK, pp. A5-53-70. IFS (Publications) Ltd, Bedford, UK.

[6]   Goto, T. et al. 1974. Precise insert operation by tactile controlled robot. *The Industrial Robot*, 1 (5) : 225-228.

[7]   Simunovic, S. 1975. Force information in assembly processes. In *Proc. 5th Int. Symp. on Industrial Robots*, September 1975, Chicago, pp.415-431. Society of Manufacturing Engineers, Dearborn, MI, USA.

[8]   Whitney, D.E. and Nevins, J.L. 1979. What is the remote center compliance and what can it do? In, *Proc. 9th Int. Symp. on Industrial Robots*, March 1979, Washington, pp.135-152. Society of Manufacturing Engineers, Dearborn, MI, USA.

A SPECIAL PURPOSE ASSEMBLY ROBOT

K. Sugimoto and S. Mohri
PERL (Hitachi Ltd), Japan

First presented in *Hitachi Review* (Vol.32, No.5, 1983). Reproduced by permission of the authors and Hitachi.

An assembly robot designed to serve as the core of a flexible, automated assembly line was developed to advance the state of factory automation technology. The robot, an articulated mechanism with six degrees of freedom, employs a new double-link drive system that extends its reach, and is controlled by a high performance controller that can handle real-time sensory feedback. The robot is also provided with a motion-orientated language that enables it to easily be taught complex movement patterns. A brief outline of this assembly robot system is given, followed by a short discussion on several technical problems that will have to be addressed in future research and development in order to come up with practical advanced assembly systems.

Current industrial robots are largely being used as semi-specialised devices for painting, spot-welding, arc-welding, and other tasks. Because these types of work require considerable skill and must normally be performed under poor conditions, it is difficult to maintain a reliable supply of experienced workers. The absence of constraints on investments in industrial robots is another reason why they are being brought into these areas of application. A third reason is that it is relatively easy to develop industrial robots dedicated to a single type of task, such as welding. In fact, painting and welding account for ony a small portion of all industrial work, which is why there exists such a strong need for the automation of assembly work in order to advance the current state of factory automation technology.

A comparison of some of the effects of the automation of assembly work using robots as opposed to the use of dedicated automation equipment reveals the following advantages of the robot system:

- A flexible assembly line capable of large variety and small batch production can be set up.
- Robots help to reduce the man-hours required to develop an automated line.
- Group control is possible on an assembly line made up of identical devices.

Fig. 1   Assembly robot – a lightweight structure was adopted, in which the driving
motor for each axis is shifted down to the base of the arm

Yet, assembly work includes a wide variety of different types of tasks, such
as insertion, caulking, screw fastening, harness assembly, and wiring;
automating these processes with robots requires advanced robot capabilities. For example, to obtain the first two effects stated above, not only the
robot, but also the system which includes peripheral equipment such as
conveyors and parts feeders, must also be versatile. Vision, touch, force, and
other sensors are of course also needed, in addition to advanced control that
feeds back from these sensors in real time. To obtain the third effect, this
type of advanced control must be achieved under a unified control system,
and the data structure and robot language for teaching movement patterns to
the robot standardised.

The assembly robot described here was developed for use as a basic
structural unit in the type of advanced assembly system discussed above. The
robot hardware and software were both designed with versatility in mind; in
particular, every effort was made to come up with a software system as
independent as possible of the hardware.

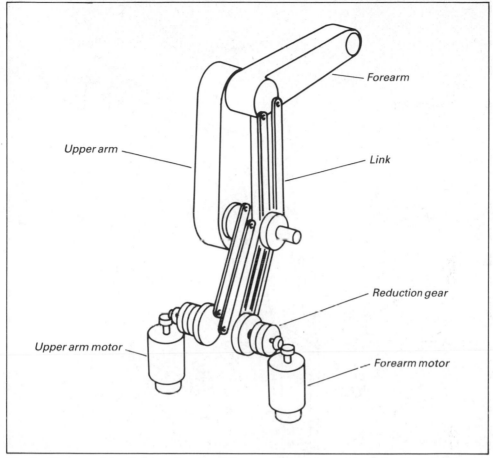

Fig. 2   The arm drive link mechanism – the reach has been extended with a double parallel link mechanism

Hardware

The robot mechanism

Locating a rigid body in three-dimensional space at a desired position and in a desired orientation requires a robot with six degrees of freedom (DOF). The robots making up a versatile assembly system must therefore have six degrees of freedom (Fig. 1). To lighten the arm on the robot, the motors were placed as far down on the base of the arm as possible and use link and chain power transmission to drive the arm and hand.

Fig. 2 illustrates the parallel link mechanism that drives the forearm and the upper arm. Standard parallel link mechanisms have an 'uncertainty position' (a point at which the direction of motion is not uniquely determined), and thus their range of motion is limited. A double parallel link mechanism was therefore adopted that combines the input link and the output link with two parallel links having a phase difference of 90°. This made it possible to pass through the uncertainty point, extending the reach of the arm.

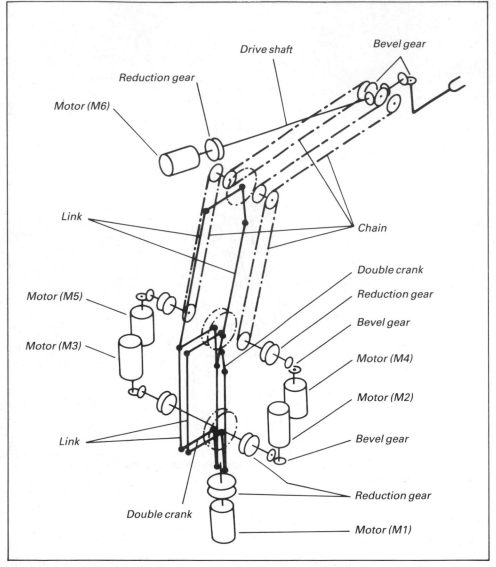

Fig. 3   Arm and wrist drive system – the drive system for a 6-DOF mechanism
consisting of a 3-DOF arm mechanism and a 3-DOF wrist mechanism

Fig. 3 outlines the power drive system for the entire mechanism, including
the wrist. The wrist has three degrees of freedom, of which two are driven
with chains by motors attached to an upper arm support mounted on a
turning table. The third is driven with a shaft by a motor attached to the
forearm. The wrist mechanism consists of bevel gears, with a means of
adjustment attached to each of the gears to eliminate backlash.

This robot mechanism is constructed such that, for use in a robot with less
than six degrees of freedom, all that needs to be done is to remove the
actuators and drive systems for the unneeded axes of movement from the
transmission system in Fig. 3.

Controller

Assembly work covers a broad range of tasks. It includes relatively easy tasks that require rapid movement, such as the transfer of a part, precision work such as the insertion of a part, and difficult tasks such as harness assembly and wiring that can be accomplished only with the use of visual sensors. In order for a robot to be able to perform these assembly tasks, more is required than just high-speed , high-precision control of the arm; the use of visual and tactile sensors is also necessary. A multiprocessor construction capable of adopting a processor configuration suited for the robot DOF and the task, and one to which a special-purpose processor for processing complex and diverse sensory data can be added, is most appropriate as a robot controller for assembly tasks. This should be given the flexibility necessary for future enhancements in robot functions. Fig. 4 shows the basic structure of a controller developed with the above in mind.

This controller consists basically of three processors, a common memory, a robot drive circuit, and other components linked together by an IEEE-P796 standard bus. The tasks of each processor are as follows. The main control processor carries out overall control of the system, external communications control, and robot language and teaching control; the calculation processor carries out coordinate transformation and the servoprocessor performs servo control. Parallel processing of each of these tasks is carried out under the control of distributed monitors (which add task startup and task termination wait functions between the processors) for multiprocessors. Communication from one processor to another is accomplished through a common memory. This type of task control system enables the free modification of the coordination between the processors and tasks without any large changes in the control software. This was designed to allow for easy functional upgrading by the addition of a sensory data processing module.

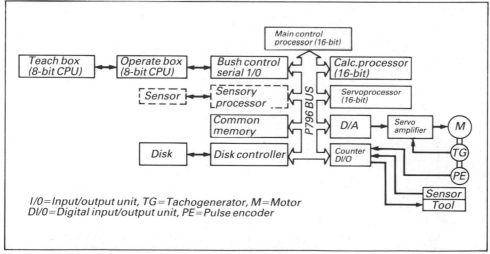

Fig. 4   Structure of controller – new functions such as a sensory processor can easily be added to this basic multi-microprocessor system, which uses a standard bus connection

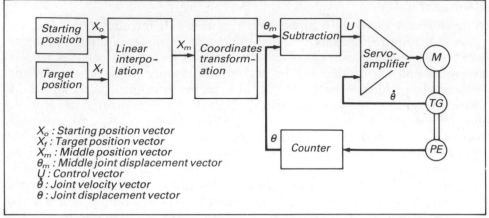

Fig. 5   *Trajectory control – the trajectory is generated by the linear interpolation of a series of midpoints between a starting position and a target position*

Control system

In industrial robots now in use, control is performed by generating a trajectory through interpolation between a number of points taught to the robot, and movement of the hand along this trajectory. A block diagram of this control system is shown in Fig 5. First, interpolation between the two points taught is performed for the designated velocity, and the midpoint determined. This midpoint serves as the target point at a corresponding sampling time of control. In order to be able to set the trajectory through interpolation, data expressing the position and orientation of the hand must be given in Cartesian coordinates. However, because control of the robot

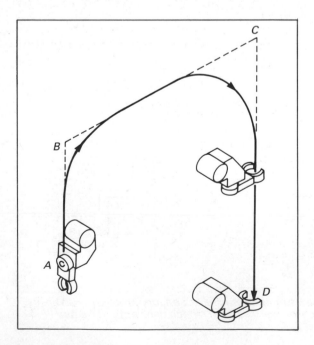

Fig. 6   *Trajectory smoothing – interpolation for a smooth trajectory is used to obtain a parabolic trajectory near points B and C*

involves the rotational displacements of each of the kinematic pairs (an element in the robot mechanism that restricts the motion between two rigid bodies, such as joints and the like), the position data for the hand given in Cartesian coordinates needs to be transformed into the displacements of the kinematic pairs. Control of the actuators that drive each of the kinematic pairs is carried out using the kinematic pair displacements thus obtained as the target points. In this method, a single interpolation and transformation of coordinates calculation is required to set one target point; the time required for this calculation accounts for most of the sampling time. Linear interpolation is normally used as the method of interpolation.

When linear interpolation is used, a polygonal trajectory is generated, with the points taught serving as the vertices, the hand is unable to pass through these points at a constant speed, making an adjustment in velocity necessary near each vertex. When applied to assembly work, the motions shown in Fig. 6 of grasping a workpiece at point A, lifting it to point B, transferring it to point C, and assembling it at point D, are common. The hand need only pass near points B and C; the speed adjustment required near those points significantly increases the cycle time of the task. To overcome this problem, the trajectory was made parabolic and thus continuous near points B and C, and provided the robot with functions for shortening the motion time[1]. The method used here generates a parabola by means of a simple linear interpolation calculation, making it appropriate for real-time calculation during robot movement.

The system just described controls trajectory generation for the hand. However, this system is not capable of handling real-time sensory feedback. To control a robot by means of sensory feedback, the instantaneous motion required for the hand must be determined from sensory signals and actual hand position data, and this instantaneous motion converted into the velocity of the kinematic pairs in the robot.

Fig. 7 is a block diagram of the control system required to do this. This diagram is an example of position control; target point data may be provided by the visual sensor in Cartesian coordinates and may change during robot

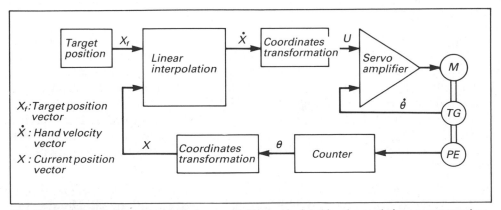

Fig. 7 Position follow-up control – control is carried out by the real-time computation of the required velocity for each kinematic pair from sensory signals indicating the actual and target positions of the hand

movement. Robot control is carried out by calculating the current position of the hand from the displacements of the kinematic pair, using the current position and target position to determine the necessary hand velocity and converting this velocity into the velocity of the kinematic pair. Robot control can handle real time sensory feedback only if all these calculations are performed in real time. The number of these calculations is much greater than for the method of trajectory control shown in Fig. 5, but processing is being carried out with the controller described in the previous section at a sampling time of 20ms.

The robot language

The robot teaching method currently used involves teaching a movement pattern and the position data at the same time. The robot is manoeuvered by means of control buttons into a specific position and the position data stored in the controller. Once the robot has been taught a sequence of motions by the repetition of this process, during playback it moves in the pattern of motions learned. Together with position data, the robot is also taught signals for interlock with external equipment and data needed for operations such as for opening and closing of the end effector. Unfortunately, teaching a robot by this method is time consuming; in addition it is difficult to make later revisions. As a result, it is becoming more common now to use procedures that involve teaching the robot information such as the sequence of motions and interlock signals in robot language, and teaching the position data separately by manipulating the robot.

Robot languages can be divided into the three levels[4]:

- Motion-orientated languages in which there is a one-to-one correspondence between instructions and the basic robot motions.
- Task-orientated languages that are independent of the robot mechanism and control system, and that express tasks during assembly by instructions such as 'grasp' or 'release' a part.
- Job-orientated languages that automatically decide where to grasp a part and where to place it with the help of a data base by describing the job that has to be done.

Languages at the task- and job-orientated level are still experimental systems under study by artificial intelligence research groups; at present, motion-orientated languages represent the mainstream in industrial robots.

The language used in this robot system is a motion-orientated language called Assembly Robot Language (ARL) that describes the movement of the robot arm and end-effector motions. ARL is capable of specifying various robot control modes, synchronisation and coordination with peripheral equipment, and also the sensory tasks. It has the following major features:

- *Continuous point input and specification of various interpolation methods.*
  In addition to using the method described earlier of specifying and linearly interpolating between two positions, ARL can also describe information on a group of points and smoothly interpolate between these.

Table 1 Outline of program grammar*

Category	Instruction	Format
Motion instruction	move	MOVE $\begin{Bmatrix} interpolation \\ method \end{Bmatrix}$ $\begin{Bmatrix} [midpoint,...] \\ point\ group \end{Bmatrix}$ $\begin{Bmatrix} target\ point \end{Bmatrix}$ [,option]
	drive on joint	DRIV joint no., $\begin{Bmatrix} /absolute\ angle/ \\ relative\ angle \end{Bmatrix}$
End-effector instruction	change	CHG end effector no.
	definition	HAND end effector no. [,parameter]
Program flow control	Repetition	FOR variable = $\begin{Bmatrix} variable \\ constant \end{Bmatrix}$, $\begin{Bmatrix} variable \\ constant \end{Bmatrix}$ ~ END
	conditional branch	IF $\begin{Bmatrix} variable \\ I/O\ variable \end{Bmatrix}$ $\left[\begin{Bmatrix} comparative\ operator \\ logic\ operator \end{Bmatrix}\right]$ $\begin{Bmatrix} variable \\ I/O\ variable \end{Bmatrix}$ THEN~ ELSE~
	repetition with conditional branch	WHILE $\begin{Bmatrix} variable \\ I/O\ variable \end{Bmatrix}$ $\left[\begin{Bmatrix} comparative\ operator \\ logic\ operator \end{Bmatrix}\right]$ $\begin{Bmatrix} variable \\ I/O\ variable \end{Bmatrix}$ ~ END
	simple branch	GOTO label
	conditional wait	WAIT I/O variable $\left[\begin{Bmatrix} comparative\ operator \\ logic\ operator \end{Bmatrix}\right]$ $\begin{Bmatrix} variable \\ I/O\ variable \end{Bmatrix}$
	subroutine branch	CALL name [(parameter [,parameter,...])]
Others	timer wait	DELAY $\begin{Bmatrix} Variable \\ Constant \end{Bmatrix}$
	substitution	$\begin{Bmatrix} variable \\ I/O\ variable \end{Bmatrix}$ = $\begin{Bmatrix} variable \\ I/O\ variable \\ constant \end{Bmatrix}$ $\left[\begin{Bmatrix} arithmetic\ operator \\ logic\ operator \end{Bmatrix}\right]$ $\begin{Bmatrix} variable \\ I/O\ variable \\ constant \end{Bmatrix}$
	position definition	position variable = $\begin{Bmatrix} positon\ variable \\ position\ constant \end{Bmatrix}$ + (DX, DY, DZ)

* This shows principal ARL grammar. The position constants given here are replaced with real values during the teaching process

Moreover, it can write on single line instructions that formerly required several lines for description, and can speed up robot movement by smooth interpolation.

- *Definition of I/O devices*. External input and output are performed via I/O ports in the controller. These ports have been named, thus enabling easy data input and output by logical names.

- *Processing of analogue data*. ARL handles external input/output consisting not only of on/off signals, but also of 8-bit analogue data. It has been provided with the ability to perform basic arithmetic and logical operations on this data. Moreover, when special processing of sensory data is required, such processing programs can easily be added, and the instructions for this processing defined as user-extended instructions.

Table 1 gives an outline of the program grammar in ARL. Not mentioned in the table is the ability to describe a variety of operations using variables, and to switch tasks based on complex conditional calculation.

An example of a program in ARL is shown in Fig. 9. The assembly task described in Fig. 8 consists of inserting Part 1 located on a magazine on Line 1 into P1 of Part 2 on a pallet on Line 2. The task is performed while checking palletising within the magazine and synchronisation with the magazine and pallet.

The external input signal I2 in Fig. 9 is preset to the I/O port address. This may also be defined in the program by an I/O definition command (IOD), as in the case of signal I1.

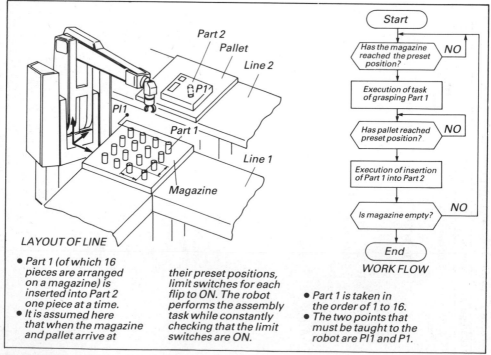

Fig. 8   *Example of assembly task that includes palletising*

```
JOB EX1
IOD I1 (0AOOH)                               I/O port address
SPEED 500                                    Default value for velocity*
WAIT I1=1                                     Wait until value for limit switch
PT=PI1                                        I1 becomes 1

FOR J=1,4
PT=PT+(0,−y₀,0)                              Position of first part in next row

FOR J=1,4
PT=PT+(x₀,0,0)                               Position of next part

MOVE A,PT+(0,0,20),S=800                      Grasp and hold Part 1
MOVE I,PT
HAND I,O
MOVE I,*+(0,0,20)                            *Indicates present position

WAIT I2=1                                     Wait until value for limit switch 12 becomes 1

MOVE A,P1+(0,0,20),S=800                      Insertion of Part 2
MOVE I,P1
HAND I,1
MOVE I,*+(0,0,20)

END

PT=PT+(−4x₀0,0)                             Position of first part in next row
END

JEND  *Default value : taken when another value is not selected as an option. (In this
example. the velocity of MOVE when a particular velocity is not specified is 500 mm/s.)
```

Fig. 9 Example of an ARL program for palletising and interlocking

Technical problems

As noted in the introduction, technology peripheral to the robot, such as that relating to parts supply and transport, is important in organising a flexible assembly system. Although this area includes a large number of technical topics that await future research and development, here we shall focus only on the robot itself.

Current methods that teach a robot tasks separately for each product are not adequate for the automation of small-batch, large-variety production, making it necessary to link up the robot with a CAD/CAM data base. When this is done, to generate the position data necessary for hand movement from the data base and position the hand at a desired point, absolute positioning precision for the robot is required. However, because the robot mechanism consists of a complex assembly of many kinematic pairs, a large number of error factors exist, making it difficult at the present time to assure adequate precision. A solution to this problem will probably require compensation for errors in the robot mechanism, and also position error compensation using sensors. The practical realisation of task-orientated languages is now being developed on an experimental basis, and the group control of robots will become possible only with this errror compensation technology.

A very important topic in robot control is real-time sensory feedback control. Although such achievements as the execution of parts fitting tasks through the feedback of data from force sensors attached to the hand, and the use of visual sensors attached to the hand to carry out position compensation through real-time feedback are now technically feasible[2,3], a vital question here is how these are to be incorporated into a versatile robot system. Real-time sensory feedback is already possible for specialised use in one type of task, but to provide a system with greater versatility, it will be necessary to successfully couple this real-time sensory feedback with standard robot control such as trajectory control, and introduce this into a robot language to permit general use. In addition, the conversion of data coordinates is necessary in real-time sensory feedback, with the number of required changes in coordinates being especially large when hand sensors are used. Because the computing time required for this coordinate transformation is the control sampling time, it will be necessary to increase the computational speed of the controller.

Concluding remarks

The assembly robot developed at Hitachi has the following features:

- A lightweight, modular 6-DOF mechanism with a large reach that can be reduced to a 4- or 5-DOF mechanism.
- An extendable controller that can also handle sensory feedback control.
- A control system that permits smooth trajectory interpolation and target point follow-up.
- A robot language that facilitates the teaching of a robot and can specify any of a variety of control modes.

With the future prospects for further enhancement of robot capabilities, the promise of lower prices and increased reliability that the ongoing technical innovations in electronics hold in store, and the improvements in the purchase and installation conditions through the cooperation of production engineers and product designers, we are sure to see flexible manufacturing systems (FMS) gain wide acceptance in assembly processes.

References

[1]   Sugimoto, K. 1983. Trajectory interpolation of a robot hand. *Bulletin JSME*, 26(213): 439-445.
[2]   Kashioka, S. et al. 1977. An approach to the integrated intelligent robot with multiple sensory feedback : Visual recognition techniques. In, *Proc. 7th Int. Symp. on Industrial Robots*, October 1977, Tokyo, pp. 531-538. Japan Industrial Robot Association, Tokyo.
[3]   Takeyasu, K. et al. 1976. Precision insertion control robot and its application. *Trans. ASME, J. Engineering for Industry*, 98B (4) : 1313-1318.
[4]   Arai, T. 1982. European trends in the development of robot languages. *Automation*, 27(4) : 27-31. (In Japanese)

YESMAN – A NEW COOPERATIVE ASSEMBLY ROBOT

P. T. Blenkinsop
PA Technology, UK
and
M. Scibor-Rylski
Prutec, UK

First presented at the 6th British Robot Association Annual Conference, 16-19 May 1983, Birmingham, UK. Reproduced by permission of the authors and the British Robot Association.

A new type of robot aimed specifically at the assembly market and which is currently at the evaluation stage is described. The market and technical factors which have influenced its design together with an overall description of the robot are presented. Some typical applications that have been investigated are illustrated and indications of future potential uses are given.

Prutec has invested in the design of a new robot product to serve the emergent market of assembly automation. The motivation for new products in this area is strong and YESMAN is aimed at solving a number of problems often associated with typical batch assembly such as high inventory, inconsistent quality levels and lost time due to line changes or poor utilisation factors.

Most robots have been rightly regarded as machine tools and their designs and manufacturing methods reflect their heavy engineering origin. YESMAN's pedigree, however, has its roots in a number of newer technologies including commercial, business and instrumentation equipment. This background has led to a robot design which will be found of interest to a wide range of industries because of the product's size, weight, cost and flexibility.

Prutec is not itself a manufacturer but a wholly owned venture investment subsidiary of the Prudential and in the project aspect of its operations is involved in product creation within certain strategic areas including robotics. The YESMAN robot is the result of a development programme at PA Technology, and Prutec as owner of the intellectual property rights is currently discussing the commercial exploitation of this and other projects.

These designs are available at an early prototype stage, with production engineering dependent upon the host company. In this way a very up-to-date level of technology can be made available with the maximum opportunity for short product lead-time. Clearly cost or cost-related topics are out of place here.

The basis on which these projects are to achieve commercial exploitation is unusual – the R & D risk has been removed but the production engineering and commercial commitment must be supplied by the host/licensee. New product initiatives in robotics are essential if the forecast 30% annual market increase is to be sustained, and if the UK is not to become simply a manufacturer of licensed designs from overseas and a supplier of applications engineering technology.

General product philosophy

To date, robots have made very little impact on the light assembly applications compared with their usage in, say, welding, paint spraying and heavy handling. This is a direct consequence of the difficulty of the manipulative skills needed in most assembly tasks – skills which come so easily to most human assembly workers. Two approaches have so far been adopted by industry to allow robots to carry out useful assembly tasks:

- Provide the robot with the visual and tactile feedback needed to mimic a human being.
- Order the surrounding environment sufficiently well (e.g. using bowl

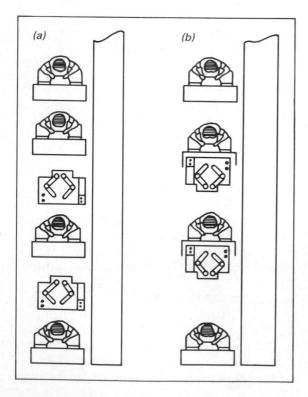

Fig.1   YESMAN (a) interspersed with manual assembly and (b) working cooperatively

feeders, bandoliered components, etc.) so that the orientative and manipulative requirements are reduced.

The first approach tends to lead to a robot which is still economically unattractive whereas the second reduces substantially the flexibility of the robot.

YESMAN has been designed around the premise that a significant increase in productivity and quality can be obtained by a combination of human and machine skills. This is achieved by designing YESMAN such that it can work safely alongside human operators.

There are several advantages of a 'mixed' assembly process. The first has already been mentioned, namely that cost is not built into the machine in trying to make it do a task for which it is inherently unsuited. Secondly, the implicit quality control that is readily achievable with manual assembly is retained. Thirdly, the combined attributes of manual and machine assembly (flexibility, precision, quality and speed) can be greater than with either alone.

It is with this concept in mind that YESMAN has been designed to operate in a number of guises. Fig. 1(a) shows an assembly line where tasks are split between robot and human and each operates independently. Fig. 1(b), however, shows a line where some workstations feature YESMAN and a human operator working cooperatively. By bringing the scale and complexity of the robot down to the level of work-aids already seen on the shop floor, it is felt that acceptance of robot aided assembly will be increased.

Factors influencing the design of the robot

In trying to gauge the market potential for YESMAN, a number of factors which influence the sale and usage of robots within the assembly market were explored. The main factors can be summarised as:

- *Design changes.* In order to permit pure robot assembly to take place it is often necessary to redesign the product to make the assembly task easier. The general quality and precision of piece-parts often has to be raised. Product redesign is often cited as a barrier to the use of robots especially where small to medium batch assembly is concerned.

- *The safety factor.* Even the smallest of current mainstream robots is potentially dangerous and therefore needs to be segregated and operated only by a limited number of personnel.

- *Portability.* Because of the need to provide electrical and other supplies, the safety factors, and the need for a datum which is often the floor, most robot installations are semi-permanent. This partially destroys the claim that they can provide a flexible manufacturing system. In small batch assembly, it is often desirable to split or rearrange the physical layout of the assembly process.

- *Quality control.* With normal assembly there is usually an element of visual and tactile inspection taking place throughout the process even if only in a very informal sense. Unless catered for by other means, this can often be lost with machine-based automation.

- *Costs.* The purchase costs of existing mainstream robots can be in the region of five to ten times the true annual cost of a manual worker with a machine life of perhaps five years. There is a tendency to use robots where very significant productivity increases can justify the investment.

- *The marketing factor.* The bulk, cost, and lack of portability of most robots usually precludes their demonstration except on the robot manufacturer's premises. Clearly if a robot can be demonstrated on the customer's shop floor, preferably carrying out a real task, then this is a significant aid to marketing.

The overall system design of YESMAN has been aimed at obviating or at least reducing these market based barriers to robot-based assembly.

In addition to the market factors, an analysis of assembly tasks was carried out to identify the technical factors which would influence the robot design.

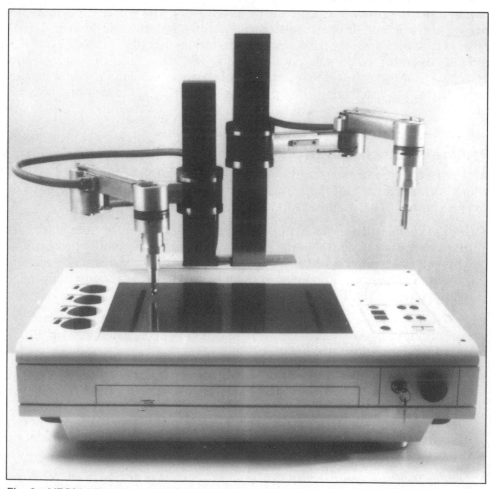

Fig. 2   YESMAN – the evaluation model

From this analysis a picture emerged of a robot capable of the following functions:

- Basic pick and place (but not bearing fit insertion).
- Self-tapping screw insertion.
- Threaded insert.
- Rivetting.
- Adhesive and sealant application.
- Adhesive tape application.
- Soldering.
- Snap connection.
- Circlip and other fastener applications.
- Adjustment of rotary or linear controls.
- Heat or ultrasonic bonding.

There is, in addition, a list of tasks which can be accomplished with more specialised tooling.

These features suggested a robot which required only a limited number of degrees of freedom *during* the assembly operation and which was required to produce forces consistent with that of the human arm provided appropriate tooling was supplied. From these basic guidelines, a relatively lightweight and mechanically simple assembly robot has emerged (Fig. 2).

Prototype

Overall design

The detail design of the machine is expected to change as more and more operating experience in a variety of tasks is gained, but the basic configuration remains as depicted in Fig. 2. The machine is essentially a workstation and comprises:

- Twin robot arms which can work either synchronously or independently.

- An integrated workbench which incorporates jig mounting features, tool holders, basic operator interface and data cartridge reader.

- An integral control computer which can be used either to control the robot or to program it.

A number of peripheral items such as a safety screen, data cartridge unit, etc., are also associated with the workstation.

The workstation has a working area of 500 × 250mm and assumes that the workpiece will have a reasonable level of vertical access. The working volume extends this area vertically by 250mm.

The working volume is served by two servo arms (although the machine can be configured as a single arm unit if required). Each arm has three servo axes powered by low-cost dc motors. The maximum normal arm load is 1kg although heavier loads can be accommodated with some loss of accuracy. The maximum arm speed is 30m/min and positional accuracy is ±0.3mm with a repeatability of ±0.1mm.

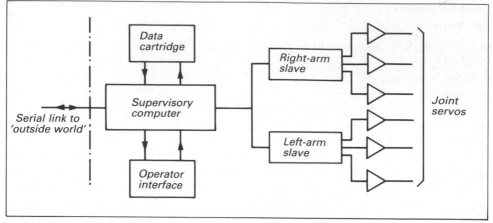

Fig. 3   Control block diagram

Tooling

A number of tool options is envisaged for the robot and the machine can accommodate a basic set including general-purpose rotary and gripper tools within sockets on the workstation. The robot is capable of locating, selecting, and securing these tools by itself. An integral electrical supply is provided but pneumatic tools can be used in addition. Special purpose tools such as recoilless hammers, glue dispensers, tape applicators, circlip pliers, soldering irons and trimmers for electrical components are all suitable for YESMAN. The mechanical assembly is sufficiently compact and lightweight that it is easily carried and repositioned by two people, making changes to the assembly line very simple.

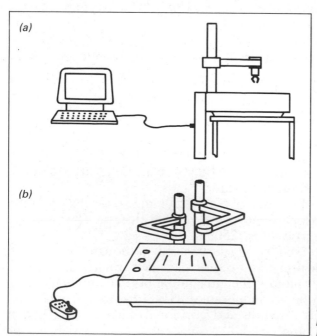

Fig. 4   (a) Off-line programming, and (b) on-line programming

Control system

The control system for YESMAN has been carefully thought out in order to minimise system cost without restricting performance unduly, and to allow future enhancements to be implemented. Fig. 3 illustrates the control system in block diagram form. At the lowest level the three servo axes for each arm are controlled by a single microcomputer which is multi-tasked. The microcomputer is part of the control loop and control is fully digital. The demand signals for these slave processors are supplied by a central supervisory microcomputer. In addition to overall supervision of the robot, the supervisory computer needs to take data relating to a task and convert it from Cartesian (x, y, z) coordinates into joint coordinates (θ, ϕ, z) in real time. To assist with this task the processor is supported by a 'maths' chip.

Program storage is on data cartridge – this being considered more reliable than disk. It is recognised that in a number of applications (e.g. ATE) YESMAN will need to interact electrically with the workpiece and a number of interface options have been designed into the control hardware.

Programming

One innovative feature of the control system is that the supervisory computer which normally handles task files at a very low level can be used to produce these files by means of a dialogue with the user using a high-level language (Fig.4(a)). In practice this means that the production supervisor is able to prepare a program cartridge for one or more robots using simply a dumb terminal and a YESMAN taken temporarily out of the assembly line. Programs produced in this way are built up using simple English language commands and absolute or relative coordinates relating to the workpiece. Tasks can be built, tried and edited in a very simple fashion and then finally stored on a data cartridge.

As an alternative to building a task from a terminal, a programming 'pod' is provided (Fig 4(b)) which enables operation of the robot via dual joystick

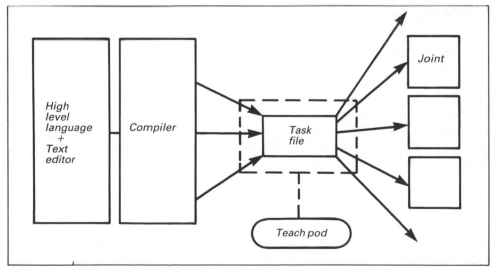

Fig. 5   Task file generation

control. This enables a number of assembly operations to be tried quickly to determine the optimal approach, and then the task can be learnt and stored on data cartridge.

The fact that task data is stored in a very primitive form is seen as a very powerful tool since it enables a task to be entered using English commands, the task to be run on the robot, and then the task parameters modified directly using the programming 'pod' (see Fig.5).

Safety

Clearly any machine which is powered and which is expected to operate in close proximity to humans requires very careful attention to operator safety. Safety has been an underlying theme of the machine development and safety features have been discussed with the UK Health and Safety Executive to facilitate compliance with the forthcoming guidelines on the use of robots. The safety features operate at a number of levels. One of the main factors is that the robot, being lightweight and designed specifically for light assembly, is unable to exert forces much in excess of a human being. Remaining dangers inherent in the robot itself are eliminated by removing trapping points by the use of end stops. The main hazards therefore are the tools which may be hot, high-speed, or be capable of considerable force or torque. The normal safety features of the machine are:

- Emergency stop button.
- Key holders are the only users.
- High-speed tools are guarded.
- Independent isolation of power on operator intrusion.

Additional safety features included or allowed for are:

- Ergonomic industrial design of controls to avoid operator confusion.
- In-built protection against anomalous operation.
- Ultrasonic or light curtain protection against intrusion into the workspace in some applications.

In summary, the design has been matched to the requirements of a wide variety of assembly and finishing operations with manufactured cost as a primary constraint, and with the development of an extremely versatile machine which minimises the need for post-sales applications engineering as a main objective.

The design philosophy of the first prototype has been to produce a practical machine with flexibility built in at every level to maximise opportunities for extending the scope of the product into new application areas.

Typical applications

Typical applications fall into three categories:

- Assembly, e.g. motor commutators, keyboard assembly, packaging (sweets/components).
- Testing, e.g. switches, ATE-probe and setting of adjustable components.

- Special features, e.g. chemical assay, chemical sampling, dangerous materials.

All cooperative robot applications need to be economically motivated. The cost per unit produced or tested, including the rework cost and the averaged cost of short-run changes, must be demonstrably less if the market is to be realised. The main cost factors are:

- Robot capital cost.
- Application tooling cost.
- Operating cost.

YESMAN is designed to reduce the first two categories considerably. No major application costs are incurred. The machine is expected to be available for the cost of one to two operator years at UK rates.

The following examples illustrate typical applications.

Switch testing

An automotive steering column switch is a self-contained assembly which performs a number of electrical switching functions. Although simple, the assemblies require 100% testing and this can only be achieved by moving the switch arms mechanically through a test sequence whilst monitoring the contacts.

An installation using YESMAN in this application consists (Fig.6.) of the

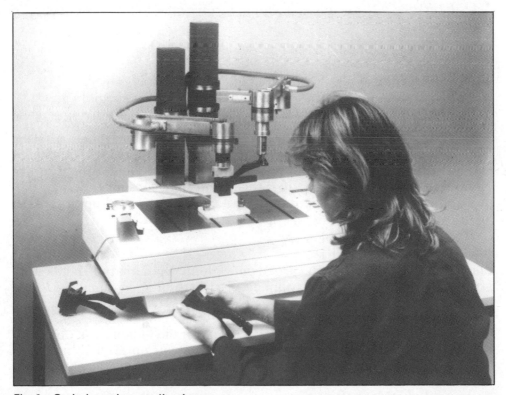

Fig. 6   Switch testing applications

robot with a test-jig to which is attached an existing continuity tester. Switches are supplied from either a conveyor or storage rack and are clamped into the test fixture. The second arm then operates the switch arms in the correct sequence.

Economic justification of the use of YESMAN stems from the savings which arise from a lower cycle time plus a better utilisation factor during the 8-hour shift. For comparison purposes the productivity is:

Manual	Cooperative
3 operators	2 robots + 1 operator
180 tests/hour	240 tests/hour
780 k tests/year	960 k tests/year
	0 missed faults

Perhaps the most significant justification cited for this application is the cost of rework when faulty switches are missed. If these units are built into cars, then the cost of stripping and replacing the switch cluster is considerable and will affect both profitability and reputation. Manual testing has been found to be fallible whereas YESMAN provides a reliable and impartial test.

Gearbox assembly

A speed reduction unit for a domestic food mixer consists of two alloy castings and four steel gears. Because of the precision fit of the gears, the task is very difficult for a robot alone.

Using YESMAN for this application involves the following sequence of activities:

- The human operator places one of the shell castings into a pallet and drops in the gears.
- The robot then applies a line of liquid gasket to the shell.
- The human then drops the upper case half onto the assembly.
- With its other arm, the robot inserts and tightens the six fixing screws.

Each of these activities could clearly be achieved by designing and installing dedicated machines. However, for historical reasons there are a number of variants of each gearbox requiring assembly, and modifications to a line based upon hard automation are difficult and time-consuming.

The economic justification for automation is again based upon replacement of two operators with one operator and a robot.

Laboratory automation

Many of the test programmes carried out within the pharmaceutical industry are both tedious and time-consuming. They are time-consuming because tests need to be initiated and then sampled at intervals, which can be as long as several hours. These tests are supervised by personnel who are sufficiently skilled that the results can be interpreted and yet the tedious task of placing reagents or test material into several tens of samples prevents them from doing other useful work. Except where a test is standardised and specific to one industry (e.g. cosmetics, tobacco, etc.), it is rare for fully automatic equipment to be installed.

Because of its scale and flexibility, YESMAN can be used within the laboratory environment to improve efficiency and working conditions. Discussions with users have suggested that an extra degree of freedom afforded by rail-mounting the robot would add an additional level of versatility and enable YESMAN to work cooperatively with a number of different items of immobile test equipment.

In this application the ability to interface the robot to other sensors and test equipment is seen as a major strength, since it is then able to interact with a data-logger to provide a record of test results and also to repeat tests, if necessary, using a different sequence.

Concluding remarks

The initial aim for the YESMAN cooperative robot was the economically viable support of light assembly roles. Selective exposure of this technology has allowed a non-comprehensive study of applications. It is now clear that the YESMAN specification is able to fulfil a much wider role than anticipated. These applications fall into two categories:

- Assembly and test application.
- Low-cost robot functions associated with other equipment – a peripheral role.

The key features which have been universally welcomed are the low cost and the freedom from application engineering overheads. The performance/accuracy cost trade-off has favoured cost. As sensors and peripherals are now becoming available and can be interfaced to robots, accuracy can be improved beyond the performance offered by the robot alone. In this way accuracy may not need to be unnecessarily and expensively bought.

A key advantage of the cooperative robot is its ability to use existing production hand tools in a semi-automatic mode. Operator skill is retained but repetition and repeatability are left to the robot. This an essential ingredient in cooperative robotics.

YESMAN now needs to be used in proving applications roles to verify the economic benefits of this new category of robotics. This, like education robots, can be an area for strong UK initiative. This requires an industry awareness of the advantages that a third generation low-cost cooperative robot system can offer and a willingness to use technology which has now been developed.

Prutec is now seeking to commercialise this technology via the licenced development and manufacture of YESMAN by suitable engineering companies. It must be noted that a production engineering phase must be carried out by the host company and that importance shall be attached to ongoing performance, cost, software and peripheral improvements.

PROGRAMMABLE AUTOMATIC ASSEMBLY STATION WITH NEW ASEA ROBOT

U. Holmqvist
ASEA Robotics, Sweden

First presented at the 14th International Symposium on Industrial Robots and 7th International Conference on Industrial Robot Technology, 2-4 October 1984, Gothenberg, Sweden. Reproduced by permission of the author and IFS (Conferences) Ltd.

The special demands placed on robot assembly equipment for medium volume production of small parts are discussed. Also described is a system developed by ASEA and how a new type of robot was designed with a different configuration called the pendulum robot.

There is no single optimal solution for automation of assembly using robots. Parameters such as production volume, size, shape and weight of the parts, number of parts included, feeding complexity, mating complexity and number of variants influence the choice of assembly system[1]. Having carried out a close analysis, ASEA decided to develop a station for medium volume production of small part assembly.

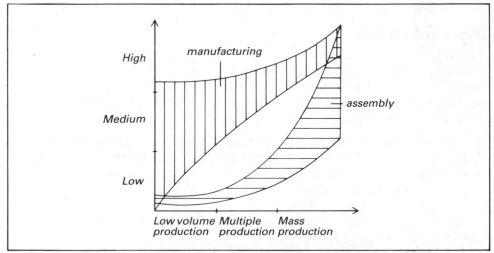

Fig. 1   Degree of automation in the field of manufacturing and assembly

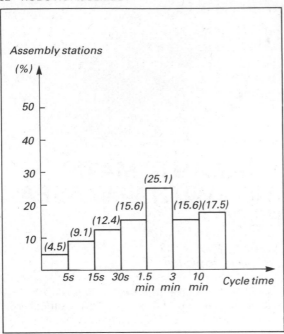

Fig. 2   Most common cycle times in assembly industry

The production volume is an important factor and implies the degree of problem complexity that has to be solved by one robot. The degree of automation in assembly is low and has so far only had an impact on high volume production (see Fig. 1). At the same time the majority of products are produced with a 30s–10min cycle time (see Fig. 2). This indicates a demand for a new type of solution for medium volume production; a system that can handle a number of parts, containing all the versatility and intelligence necessary to put together parts that are produced in such low numbers that optimising their design for automated assembly does not pay off.

System demands

The demands for high flexibility, batch production and short installation time imply the need for standardisation. It must be possible to solve the following problems automatically with the use of an assembly system including only one robot[2] (each separate problem has to be standardised to achieve a total standard solution):

- *Part presentation.* Many parts can be handled by conventional standard feeding systems like bowl feeders, hoppers, etc. Other parts, however, must be handled separately because of the risk of wear or the shape of the parts makes feeding impossible. It is important to make sure that the system has a magazine large enough to store parts for sufficient time and that all manual loading is done in bulk. Otherwise the station will demand too much manual supervision.

- *Part mating.* To be able to put a low number of components together it is normally sufficient to use a robot with three or four degrees of freedom and one special purpose gripper. The low production volume, however, makes it necessary to handle at least six different parts that usually cannot easily be inserted from one direction. The need for flexibility and sensor capability also indicates the need for a higher number of robot axes. The robot also spends most of its time moving between magazines and fixtures which is why the speed and acceleration are critical. One gripper is no longer sufficient. A gripper system that can handle many parts is necessary. This can be done in two ways: either the robot changes to the correct gripper, which will use important cycle time, or it can hold all the necessary grippers, which will cost pay-load.

- *Fastening*. The product that has been put together must be fastened to make it possible to transport it in a non-synchronised way. This is normally done with a tool that is handled by the robot. To avoid the tool interfering with the mating process, the robot has to have the capability of fetching and leaving the tool in a simple and quick way. The fasteners also have to be automatically fed to the tool to speed up the fastening process.

- *Control of assembled product*. An advantage with automatic assembly is that it is possible to control the quality of the product directly after it has been assembled. Therefore no defective products have to leave the assembly station. The checks are done using different types of sensors.

- *Product transport*. The assembled product has to be transported away from the station. This can be done with some type of conveyor or magazine.

Existing robot configurations

There are three main robot configurations used today in assembly applications (see Fig. 3):

- Rectangular coordinate types.
- Horizontally jointed types (SCARA).
- Universal types (revolute).

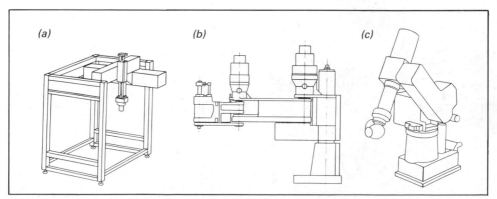

Fig. 3   The three most used types of robot configurations for assembly: (a) rectangular coordinate type, (b) horizontally jointed type (SCARA), and (c) universal type

Rectangular coordinate type

This type is very useful as a 3- or 4-axes robot. It can be built with very high accuracy, has simple control algorithms and can easily be calibrated to the surrounding world. It is well suited for simple but delicate micro-assembly applications. It has, however, limitations in acceleration due to the large overhang the arm itself represents. If the concept is extended to a 6-axes configuration, the advantages are lost because it has no longer any simple control algorithms which will affect the accuracy as well as the calibration capability.

Horizontally jointed type (SCARA)

This type is designed for three or four degrees of freedom, to insert parts vertically, fast and accurately. The design has become popular and seems to be a good solution for high production volumes, with products well designed for automatic assembly and a limited number of parts are handled by each robot. They have, however, simple software with poor sensor capability and a limited working area. If the concept was extended to a 6-axes configuration, the basic idea would be lost.

Universal type (revolute)

This type is designed for process applications like arc welding. It is very versatile but has four major limitations when used for assembly: their complex design and control algorithms make it difficult to obtain sufficient accuracy; the configuration is such that large mass forces are created which

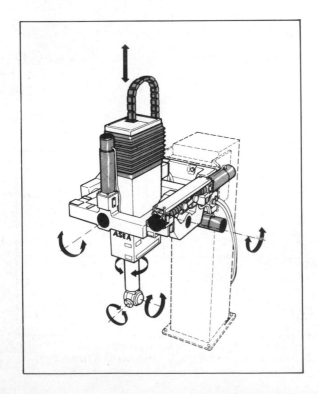

Fig. 4   The pedulum robot designed for small part assembly

limits the acceleration; the work envelope is often large but in the vertical direction which is of less interest for small part assembly; and the base turn motion creates a poor station layout and slows down the cycle time.

The pendulum robot concept

After considering advantages and disadvantages with the three types of robot configurations, it was decided that none of them were optimal for assembly of small parts in medium production volumes. A new robot concept combining the advantages of the existing ones was therefore designed.

This resulted in the *pendulum robot* concept. It consists of a hanging arm with three degrees of freedom containing a vertical stroke. The wrist contains two degrees of freedom and can be rotated. The whole arm pivots in two directions thus giving the robot its name (see Fig. 4).

This robot configuration has the following advantages when used for small part assembly:

- *Very high acceleration and speed*. The mass forces are minimised which is why all the motor power can be used for acceleration and the servo control system can run with a much higher amplification.

- *High accuracy*. The control algorithms become simple.

- *High versatility*. The fact that the robot has six degrees of freedom combined with the hanging arm means that the robot can perform any required motion.

- *Compact working area*. The concept makes it possible to utilise the working area from all directions achieving a total layout which is very compact (see Fig. 5).

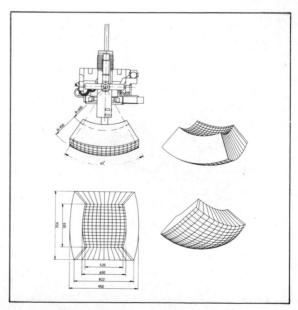

Fig. 5   Working area of pendulum robot

Existing gripper systems

Currently there are two basic systems used for handling a number of parts including tools: gripper changing systems and indexing multigripper systems.

Gripper changing system

The robot is equipped with a tool changer unit. This makes it possible for the robot to select a new gripper tool every time it has to pick a new part. The system becomes very flexible but creates a fair amount of lost time in changing tools. This can often be critical for the pay-off of the system. One way of minimising the lost time is to build many products in parallel. This means, however, high cost fixtures and normally a less flexible system.

Indexing multigripper system

The other solution is for the robot to always carry the necessary grippers and tools. This will limit the pay-load but in small part assembly the weight of the components is normally very low. There is, however, another problem due to limitations in the robot software. The robot can only work with one gripper at a time and it has to be parallel with or perpendicular to the end plate. If this is not the case, there is no software capability to turn the gripper to the correct pick-up position or insertion when programming.

New gripper system

Considering the possibilities, cycle time was found to be a critical factor, which is why it was decided to develop a gripper system with capability for

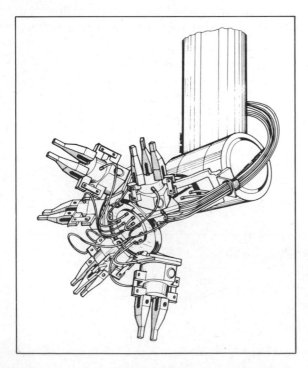

*Fig. 6   Multiple gripper system –
motion control by robot software*

Fig. 7   Screw-driving system for robot handling

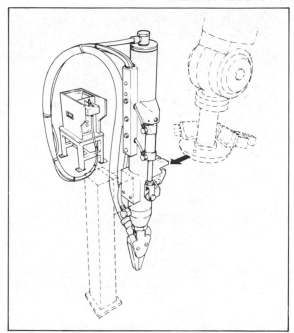

the robot to control up to six individual double-acting grippers simultaneously. It is mounted on the wrist end plate and can turn ± one revolution (see Fig. 6). To limit the gripper configurations so that only one gripper in the correct position could be used at a time was not wanted. Therefore the robot software was extended to include the standard gripper system. This means that each individual gripper can have its own TCP even if it is not mounted parallel to or perpendicular to the wrist end-plate. The system can also be used for tool handling. The standard screw-driving system (see Fig. 7) can for instance be picked up by the robot when it is fitted with the new gripper system (see Fig. 8).

Flexible assembly station concept

The station is built up around the new pendulum robot with its unique gripper system. Other standard modules are:

- Screw-driving system with automatic feeding of screws and handled by the robot.
- ASEA Vision with the possiblity of mounting the camera on the robot. It can be used to identify incoming parts, guide the robot to the correct pick-up position and control the finished product. The vision system is integrated into the robot control system and is programmed with the same man–machine–communication system.
- Standardised magazine that can supply the robot with parts on trays for 1–8 hours of continuous production. It has its own control system which makes the installation very simple.
- Other standard equipment like bowl feeders, hoppers, conveyor belts, sensors, etc.

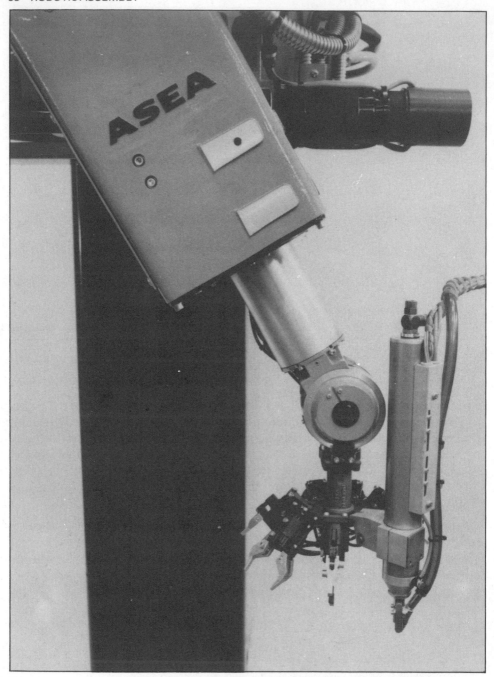

Fig. 8   Pendulum robot equipped with multigripper system and screw driver

Together these modules make it possible to present the parts, assemble the parts, fasten the parts together, control the assembled product, and transport the finished product, in a fully automatic way with a minimum of manual supervision (see Fig. 9).

Fig. 9   Flexible assembly station

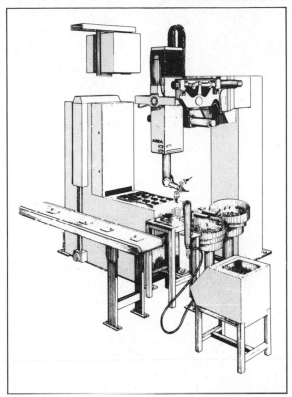

References

[1] Witte, K. W. 1982. Flexible automatisierte Montagenotwendigkeit, Vorausset-
     zungen, Lösungen. *Industrie Anzeiger*, 11 : 81–85.
[2] Skoog, H. and Holmqvist, U. 1983. Matching the equipment to the job.
     Assembly Automation, 11(4) : 211–214.

2
Robot Assembly Systems and Applications

The essence of successful robotic assembly is perhaps applications expertise. In this collection of six quite different case studies a variety of assembly requirements are studied on an international basis.

STATE-OF-THE-ART OF AUTOMATIC ASSEMBLY IN JAPAN

H. Makino
Yamanashi University, Japan
and
K. Yamafuji
University of Electro-communications, Japan

First presented at the 5th International Conference on Assembly Automation, 22-24 May 1984, Paris. Reproduced by permission of the authors and IFS (Conferences) Ltd.

The state-of-the-art of automatic assembly in Japan is reviewed. In both fixed and flexible automation Japan has much experience. Generally, mass production products are assembled in high-speed automatic assembly machines using cam-operated mechanisms and dedicated parts feeders. Recently a major change has taken place in manufacturing processes, i.e. that of flexible automation. Not only machining but also assembly processes are being automated with robots and other flexible equipment. Examples are shown and the reason why flexibility is needed is discussed

Japan is now one of the most advanced countries in the field of manufacturing automation. At the end of World War II, economic activity in Japan, like the country itself, was small. After 1950 some heavy and light industries began to 'wake up', such as the iron, chemical and textile industries, followed by the electrical and home appliances industries. At that time production was by the so-called 'man-sea tactics', in that the sea (the factory) was filled by many human workers. Consequently, employment grew rapidly and the quantity of production became larger.

In the 1960s, production volume increased sharply, especially in the automobile and electrical industries, and there was a shortage of labour. Manufacturing automation was first introduced in machining processes and then in assembly operations. Through the individual efforts of the workforce (and not by governmental policy), the quality of Japanese products was improved remarkably.

In the 1970s, with the rapid growth of electronic and computer technology, high-technology products became popular (such as TVs, VTRs, audio amplifiers, cassette tape recorders, and later microcomputers and peripherals), while small cars and motor cycles established their status firmly in the world market. These large quantity and excellent quality of products are supported by the development of technology on production processes, production control, and manufacturing automation. For instance, electric discharge machining made progress in Japan, as well as quality control techniques, and numerical control methods were developed powerfully.

The 1970s also saw much progress in automatic assembly. The conventional method of assembly automation was established for large volume and small variety products. Many kinds of feeding and transfer devices were developed and high speed cam-operated systems became increasingly popular. At that time, the problem of automation in small-lot and large variety production remained unsolved.

Now, in the 1980s, a remarkable revolution in assembly automation is taking place. That is, flexible assembly using robots. The SCARA robots and the other types of assembly robots were developed and introduced in practical assembly processes. More than fifteen flexible assembly lines are already installed in advanced assembly factories – some examples are described later.

There is a tendency for manufacturers to automate assembly lines using flexible systems rather than fixed ones, since product design and model changes occur so often. Almost certainly, the assembly robot is the nucleus of the flexible assembly system.

Dedicated automatic assembly machines

Dedicated or special-purpose automatic assembly machines are used for many types of product. Typical examples are outlined below.

Car parts assembly

In the early 1940s the first transfer machine for machining was manufactured in Japan by Mitsubishi Heavy Industries Ltd for machining aircraft engines. After World War II this kind of special-purpose machine was not made until the Japanese automotive industry revitalised the needs for mass production in the late 1950s.

Fig. 1   The four-cylinder engine block assembly machine manufactured by Hitachi Seiki Ltd

Fig. 2   Fully automated assembly machine for felt pens

The introduction of special-purpose transfer machines into production lines and the progress of the mass production technology aroused a new demand; that is, the introduction of automatic assembly machines. Since the 1960s, the Japanese automotive industry has introduced various kinds of machines for automatic assembly of such products as cylinder heads, engine blocks, pistons and connecting rods, alternators, spark plugs, etc.

Fig. 1 shows the four-cylinder engine block assembly machine manufactured by Hitachi Sciki Ltd. The main features of this machine are: partial mounting of the cylinder block; synchronous transfer of bearing caps which are selected from a feeder according to the type of cylinder block using a fixed cycle transfer mechanism; and a manual mounting station for bearing caps. This machine can automatically assemble 28 pieces of 15 different kinds of parts on a cylinder block. In addition to automatic press-fitting there is one type of pin which has to be inserted manually into the cylinder, and all other parts are automatically supplied from hopper feeders, magazines or conveyors.

Felt pen assembly

Pentel has developed[1] high-speed special purpose machines for automatic assembly of felt-tip and ball-point pens (Fig. 2). The production rate of one machine is 200 pieces per minute, or about 1.2 million per month. The production line is operated by one human worker.

The workstation is composed of intermittent and continuous motion units, whose selection depends on the type of workpiece and assembly process. These units are mainly constructed by a train of constant-speed turning tables, as shown in Fig. 3.

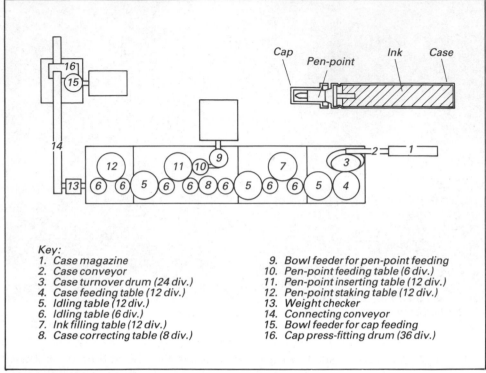

Fig. 3   Pentel felt-tip pen assembly machine layout

Key:
1. Case magazine
2. Case conveyor
3. Case turnover drum (24 div.)
4. Case feeding table (12 div.)
5. Idling table (12 div.)
6. Idling table (6 div.)
7. Ink filling table (12 div.)
8. Case correcting table (8 div.)
9. Bowl feeder for pen-point feeding
10. Pen-point feeding table (6 div.)
11. Pen-point inserting table (12 div.)
12. Pen-point staking table (12 div.)
13. Weight checker
14. Connecting conveyor
15. Bowl feeder for cap feeding
16. Cap press-fitting drum (36 div.)

One of the important features is non-stop or continuous feeding and assembly of parts. To achieve this, numerous technical developments were made. Parts feeding is automated using chutes and conveyors; the vibratory bowl feeder had to meet the feeding rate of 200 pieces per minute.

Watch assembly

In the late 1960s the Seiko Group developed a revolutionary computer-controlled fully automatic assembly system called 'System A' for watch assembly. It is said that the success of this system depends on: product redesign, improving the machining accuracy of each workpiece to $\pm 5\mu m$, assembly without adjusting and inspection, and computer control of the whole system and production schedule.

However, the defect of this system was that it was rather lacking in flexibility for the change of products, and retooling of the system required much machine down-time. Therefore the system demonstrated its merits only to large-batch products.

To overcome the defect, Suwa Seiko Ltd, a branch company of the Seiko Group, developed the 'LA-10' machine, which can produce a variety of small-lot products[2]. The LA-10 system has ten linear assembly stations as a unit whose transfer mechanism is driven synchronously (Fig. 4). On each station various working modules can be installed, such as assembly, turn-over, pick-and-place, inspection, etc.

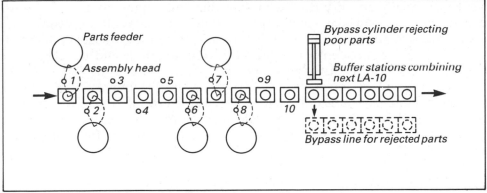

Fig. 4   Layout of LA-10 machine for watch assembly

Fig. 5 illustrates a cam operated pick-and-place module. Although it is very simple, it can assemble almost all workpieces of watch movement mechanisms. The machine can be extended by adding other LA-10 units.

Flexible assembly using robots

Already over 10,000 assembly robots have been produced and are working in Japan. Among them the SCARA type robots count for 70%.

The SCARA, Selective Compliance Assembly Robot Arm [3,4] was developed by Professor Makino with the support of several companies (see page 13). The robot basically has four axes and is designed for assembly

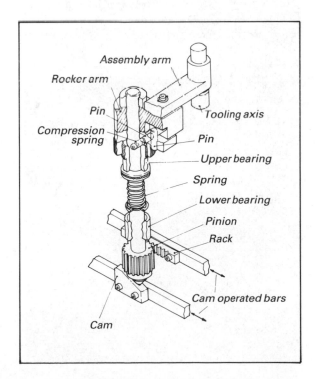

Fig. 5   Cam operated pick-and-place module

work, especially for inserting and screwing operations. It has a robust construction and can bear a high load, move quickly and position accurately.

Nitto Seiko, Sankyo Seiki, Pentel and NEC were the earliest suppliers of the robot and were members of the SCARA research and development consortium. Hirata, Mitsubishi Electric, Matsushita and Suwa Seiko followed to produce under license. Exports as well as domestic sales have started, yet the amount is not so large. It is well known that Sankyo supplies the robot to IBM as an OEM, and IBM sells the robot under its own name.

The application of the assembly robot is classified into the four categories discussed below.

Flexible loader/unloader

A robot is used as a flexible loader for some types of machines or as an unloader from a machine or a conveyor. Fig. 6 shows the Picmat-SCARA, made by Nitto Seiko, being used for loading ICs into a soldering bath. In this case the function of flexible positioning is utilised. By using the robot, feeding devices are simplified; if a fixed pick-and-place unit was used, then some kind of step-feed apparatus would be needed.

To accompany the robot, a 'tray-feed' system is becoming popular. The parts tray is a planar magazine of a two-dimensional parts presentation device. It is used for some delicate and complex parts which could not be fed by conventional hopper feeders.

Fig. 6   Picmat-SCARA robot loads ICs into a dip soldering bath

Fig. 7 Seiko robot unloads and palletises watch parts on a tray

Palletising, the operation of unloading workpieces from a point to other multiple points in a matrix, and depalletising are easily performed by robot. Depalletising is done at each assembly station to feed parts with tray feeders, and palletising is usually done at the end of the conveyor to unload finished products on a tray or in a box. Fig. 7 shows a Seiko robot, made by Suwa Seiko, unloading watch movement subassemblies onto a tray. A turn-over unit is used for reversing the face of the workpiece before unloading.

Stand-alone assembly machine

A robot may be used as a single, independent assembly machine. Fig. 8 shows the Picmat-SCARA assembling plastic retainers into interior trim panels of car doors. Each retainer is inserted into a hole and then slid in a predetermined direction. The work needs a large thrust force of up to 200N.

Assembly centre

The assembly centre is a new concept where several kinds of parts are assembled in a station without transfer[5]. It is suited to small-lot production. One or two, or sometimes more, robots are used with multiple fingers and tools. There are three methods of tool change: multiple fixed tools, turret, and ATC (automatic tool change).

Fig. 9 shows a Seiko robot inserting index chips onto the dial plate of a watch. Three types of indexes are fed individually from each vibratory feeder and the robot picks them up by a vacuum chuck and sets them onto the plate. The chuck has three hollows to fit the shape of each part.

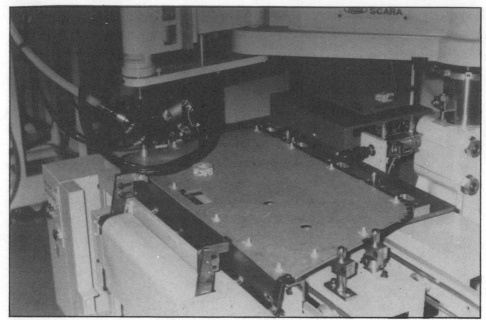

Fig. 8   Picmat-SCARA assembling plastic retainers into an interior trim panel of a car door

Flexible assembly lines with robotic stations

For large volume and large variety production, flexible assembly lines with robots are introduced.

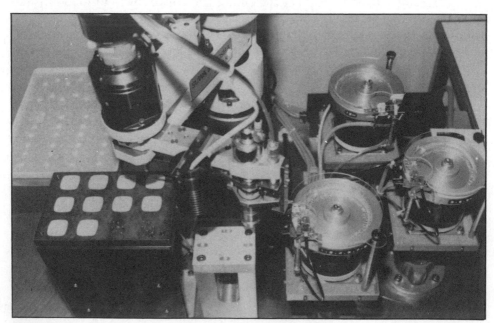

Fig. 9   Seiko robot inserting index chips onto watch dial plates

Table 1 Robotic assembly lines in Japan

Company	Plant	Products	No. of robots	Type of robots	Cycle time(s)	Installation date
Sony Taron	Omigawa	casette tape recorders	6	Cartesian coordinate	28	March '81
Pioneer	Ohmori	printed circuit boards	8	SCARA	36	June '81
Toshiba	Nagoya	electric fans	17	12 SCARA 5 Cartesian	12	February '82
	Fukaya	TV/pcbs	9	SCARA	20	December '82
Matsushita Denko	Hikone	electric shavers	3	SCARA	8	May '82
Oki Electric	Tohoku	dot printers	13	cylindrical	40	July '82
	Takasaki	dot printers	49	cylindrical	60	November '82
Hitachi	Tokai	VTR mechanisms	13	11 SCARA 2 Cartesian	8	October '82
Epson	Hirooka	dot printer heads	8	SCARA	40	October '82
Fanuc	Fuji	servo motors	60	cylindrical	?	December '82
Toyota Koki	Kariya	vane pumps	15	9 articulate 4 SCARA 2 Cartesian	45	April '83
Pentel	Ibaragi	tube paint packaging	6	SCARA	15	April '83
Sankyo	Kofu	cassette tape recorders	12	11 SCARA 1 Cartesian	18	May '83
Yamaha Motors	Iwata	motor cycle engines	20	SCARA	18	June '83
Mitsubishi	Kani	circuit breakers	4	SCARA	?	November '83

Table 1 lists the typical flexible lines in Japan which use assembly robots. The first one, developed by Sony, does not use assembly robots in the strict sense[6,7]. The line has six NC assembly centres which have normal NC tables and multiple assembly tools fixed to the frame of the machine. Both sets of parts and assembly fixtures are conveyed on a pallet to the assembly centres. Parts are stacked manually at the beginning of the conveyor.

Pioneer Ltd introduced the first model of the SCARA robot made by Sankyo to its printed circuit board assembly line (Fig. 10) at the Ohmori Plant[8]. The efficiency of assembling in this line is 99.7% as a result of the selective compliance facility. The remaining 0.3% is caused by the instability of feeding.

Toshiba developed its own robot which has the mixed configuration of PUMA and SCARA (Fig. 11)[9]. Twelve of these robots were installed together with five Cartesian coordinate types for the assembly of electric fans. The first two axes are fixed after setting, and then the robot moves as a SCARA. This line reduced the human workforce from 64 to 5 and the retooling time from 40 to 3 minutes. The introduction of the robot line not only gave flexibility for product change (in this case 24 different models), but also flexibility for product volume.

Fig. 10   First model of the Sankyo Skilam robot assembles pcbs at Pioneer Ltd

Matsushita Denko Ltd uses three robots in the assembly of electric shavers[10]. Two robots are used for palletising and depalletising motions and the third one is used for delicate assembly work. Fig. 12 shows the third robot at work. The PUHA robot, made by Pentel Ltd, picks up the motor and inserts its shaft into the hole of a plastic part with the aid of a visual processor.

Fig. 11   Toshiba robots soldering the wiring of electric fans

Fig. 12 Pentel PUHA assembles electric shaver parts with the aid of a TV camera

Yamaha Motor Cycle Ltd is one of the earliest user of assembly robots. Development of its own CAME robot began in 1979. Now the company has already installed some 80 robots in its manufacturing and assembly processes. Fig. 13 shows the typical assembly station for motor cycle engines using the CAME robot. Also shown is a multilayer vibratory feeder.

Fig. 13 Motor cycle engines are assembled in the Yamaha line by CAME robots with a multilayered parts feeder

Large scale factory automation is now rapidly progressing. The Hitachi Tokai[11] and Oki Takasaki plants[12] are two famous advanced factories. They utilise robots, automatic guided vehicles, as well as hard or fixed assembly stations, and automate the whole floor of the assembly factory. The complete assembly line is computer controlled with the computerised warehouse.

The features of these advanced flexible lines are characterised as follows:

- The products are of high technology home appliances or office automation devices, such as VTRs, cassette tape recorders and dot printers. The price of the product is rather high. The size of the product is larger than one which was assembled by the conventional dedicated machines. The number of parts is also larger than usual, between 20 and 300.
- These products will suffer frequent model changes. Progress in the product technology is very rapid and sometimes as a result of the change of customer demands. Nevertheless they need mass production.
- The manufacturer in most cases is a large manufacturer of electric and home appliances. Competition between companies is very fierce and they must lower the cost. They have fair experience in production engineering, production control and manufacturing automation.
- These lines are introduced as the model line of flexible automation, and the next generation's automation is tried. The suitable product for flexible

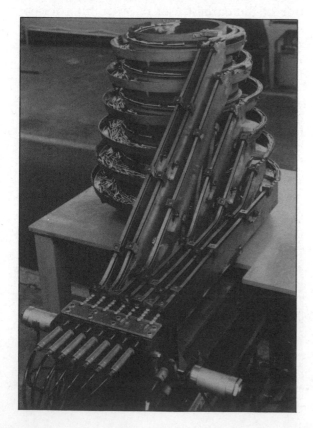

Fig. 14   Multilayered vibratory bowl feeder

automation is carefully chosen, and detailed planning is made. Product design is changed from the outset.

- Often these manufacturers are suppliers as well as customers of factory automation products. They would like to sell their own robots and computers. These lines sometimes become a show room for their factory automation products.

Parts feeding and transfer devices

Yamaha Motor Ltd together with Professor Yokoyama of Kanazawa University developed a multilayered vibratory bowl feeder in 1978[13]. Fig. 14 shows this feeder on which a plastic cover is usually installed for preventing noise. Its features are as follows: seven stages of bowls, therefore it can supply seven kinds of parts; it occupies a very small space, only 1/5 to 1/7 of that of conventional feeders; vibration is balanced between the upper four layers and lower three layers, and reaction forces are cancelled; high cost-performance; and outside tooling, i.e. tooling of this feeder can be conducted outside the bowls except for the top one.

One of the important features of this feeder is ejection of workpieces using newly developed lateral linear feeders closely attached to the exit of the track of the multilayered feeder.

FIT Corp. developed a non-vibratory bowl feeder[14] which is composed of inner and outer discs rotating in opposite directions (Fig. 15). Maximum rotating speeds of the discs are usually between 10 and 15m/s; 100m/s in extreme cases. It can supply more than 4000 parts of 10mm length per minute.

Motron Seiko Ltd produced a new type of vibratory feeder which is vibrated mechanically by a cam mechanism. The use of a cam, driven by a variable speed motor, enables vibrating frequency to be changed easily.

Professor Makino developed a new type of indexing mechanism called

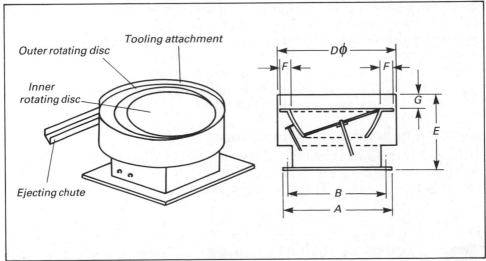

Fig. 15   Non-vibratory bowl feeder

Fig. 16   Tri-Cam indexing unit

'Tri-Cam' with three cams on the driver shaft. He has studied several variations of the parallel indexing cam. There are three types of parallel cams; namely, external and internal circular indexing and a linear one. The Tri-Cam is made by Sanko Senzai Ltd (Fig. 16).

Recently, the introduction of linear motors as drive equipment in the base machine of assembly lines is increasing. Yamaha Motor Ltd uses linear motors in its motor cycle assembly line in which motor cycle body frames are transferred by linear motors. Toyoda Machine Tool Ltd uses linear motors as a pallet transfer system. Pallets are transferred with a high initial speed of 40m/min, then decelerated by the linear motor of the next station and stopped with a stopper[15]. Return speed of the pallet reaches about 120m/min. With the use of a linear motor transfer system it is easy to construct free-flow type transfer systems.

Electronic parts insertion machines

Although insertion machines for electronic parts are increasing in speed year by year, the fastest insertion cycle time is only 0.2s per part. In order to increase this it may be better to adopt multitooled heads or other similar methods.

Fuji Industry Ltd made a breakthrough in 1976. It developed a multi-insertion machine (Fig. 17) for discrete type electronic parts[16]. About 424 pieces of electronic parts, at maximum, are inserted simultaneously into a printed circuit board through a jig block in about 7s.

Fig. 17   Multi-insertion machine

Electronic parts such as resistors, diodes and ceramic condensers are suitable for this machine. These components are fed through a jig block in which parts feeding tubes (maximum 424 tubes) are set as vertical chutes. The most important feature of the machine is easy changeability of products. Retooling of the jig block is carried out by quick exchange of the jig block just like the exchange of a die set of a press machine. Tooling is needed to arrange and position the vertical tubes according to the printed pattern on the pcb.

Some other companies are also challenging with these kinds of multi-insertion methods. Sony Corp. uses a multi-holed template for discrete electronic component mounting. Apart from component mounting, this technique finds another application. A new mounting method made by Ikeda Automatic Machine Ltd is unique and smart. In this method parts are settled into the holes in a template by vibration [17]. In order to assemble part A and part B, a template A containing part A, and a template B containing part B, are stacked, and then through press-fitting of part A with a multi-plunger pusher into part B, parts A and B are assembled together.

This is in fact a multi-mounting and simultaneous multi-assembling method. On one side of the template systematic holes are drilled according to the part patterns, about 3000 – 4000 depending on the type and size of the product.

The advantages of this method are:

- Parts settling and assembly rate can reach up to 10,000 – 100,000 parts per hour.
- Change of product can be carried out by exchanging templates.
- No jamming is caused by deformed workpieces.
- Easy and accurate positioning of parts using templates.
- Easy assembling by pressing and extruding parts settled into the template.

Recently CKD Corp. has developed an electronic components mounting system which can be applicable to almost all electronic components mounting and various processing on the pcb. Each unit is designed as a module, such as a chip placer, IC placer, loader and unloader of the board, dispenser, screen printer, parts feeder, reflow soldering unit, parts inserter, inspection unit, conveyor, magazine, etc. By arranging these modules a flexible manufacturing system for electronic component mounting can be constructed. This system is controlled by microcomputer.

Assemblability evaluation and product design improvement

Ease of production and assembly

Through Japan's twenty years experience in assembly automation it has been found that the most important ways for the improvement of productivity are in the following three categories:

- Improvement of design for ease of production.
- Mechanisation and automation of production.
- Improvement of working processes and increase of production efficiency.

Among these categories the importance of the improvement of design for production is still not recognised properly. This means the simplification of manufacturing processes, simplification of product design, and reduction of the number of parts to be assembled. Improvement of design reduces the scale of the production system, consequently the investment for the equipment is also reduced. Because simple and well-designed production equipment is available, high reliability and production efficiency of the system are attained, as well as high product quality.

Hitachi Ltd has developed an assemblability evaluation method (AEM)[11] for the evaluation of the products and parts design to obtain a

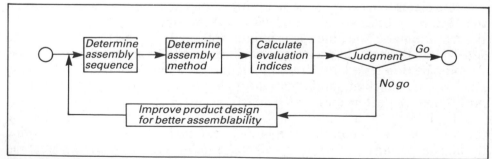

Fig. 18   Assemblability evaluation procedure

high degree of assembly automation. Fig. 18 shows the flow of evaluation and improvement of product design using AEM. The objective of AEM is that product design is evaluated from the point of view of assemblability in the early stage of designing. Final calculation of evaluated mark and estimation of assembly cost of the improved design of product are carried out in detail after the redesign has been performed.

By the quantitative evaluation of the effect of improved design, the improving activities can go smoothly. The result of an evaluation between newly designed video tape recorder mechanisms and conventional ones using AEM showed that the total number of parts are reduced from 460 to 379 (17% reduction) and the assemblability evaluation mark is increased from 63 to 73 (full mark is 100). Hitachi Ltd emphasises that the accuracy of the evaluation is within ±15%.

How to improve product design

The ways for improving the product design can be summarised as follows:

- Combination of the functions of component parts.
- Improvement of product design for easy feeding and handling.
- Increase of assemblability (assembly from one direction, adding chamfer, etc.).
- Stable resting form of parts.
- Selection of easy fastening means.
- Simultaneous parts forming an assembly in the assembly station.
- Continuous forming and feeding of parts (taped carrier, film carrier, etc.).
- Dividing into subassemblies is recommended.
- Consideration to get easy positioning of parts.
- Standardisation of parts.

References

[1] Mizuno, M. 1982. Development of fully automatic continuous assembly machine for Pentel sign-pen. In, *49th Tech. Meeting of the Committee for Automatic Assembly*, pp.1-9 (In Japanese). Japan Society of Precision Engineering, Tokyo.

[2] Yamazaki, Y. 1983. Automatic assembly of watches. *Advancement of Automation*, 12 (5): 9-15.

[3] Makino, H. and Furuya, N. 1980. Selective compliance assembly robot arm. In, *Proc. 1st Int. Conf. on Assembly Automation*, pp.77-86. IFS (Publications) Ltd, Bedford, UK.

[4] Makino, H. and Furuya, N. 1981. SCARA robot and its family. In, *Proc. 3rd Int. Conf. on Assembly Automation*, pp.433-444. IFS (Publications) Ltd, Bedford, UK.

[5] Makino, H. 1971. An experimental approach to NC assembly center. In, *Proc. Int. Conf. on Production Engineering*, pp.486-491. Japan Society of Precision Engineering, Tokyo.

[6] Taniguchi, N. 1983. Present state of the art of system design on automated assembly in Japan. In, *Proc. 4th Int. Conf. on Assembly Automation*, pp.1-14. IFS (Publications) Ltd, Bedford, UK.

[7] Sasaki, N. and Kon, T. 1982. Flexible assembly center system. In, *49th Tech. Meeting of the Committee for Automatic Assembly*, pp.11-20. Japan Society of Precision Engineering, Tokyo.

[8] Nakano, Y. 1982. Application of robots to insertion of electronic parts. Preprint for, *Tech. Meeting of the Committee for Automatic Assembly*, No. 82-5, pp. 1-6 (In Japanese). Japan Society of Precision Engineering, Tokyo.

[9] Flexible assembly line for electric fans at Toshiba Nagoya plant. *Nikkei Mechanical*, July 1982: 42-52 (In Japanese).

[10] FMS case study. *Nikkei Mechanical*, August 1982: 82-105 (In Japanese).

[11] Ohashi, T. et al. 1983. The automatic assembly line for VTR mechanisms. In, *Proc. 4th Int. Conf. on Assembly Automation*, pp.350-361. IFS (Publications) Ltd, Bedford, UK.

[12] Tanimoto, H. 1983. Small dot-printer assembly line. In, *51st Tech. Meeting of the Committee for Automatic Assembly*, pp.29-53 (In Japanese). Japan Society of Precision Engineering, Tokyo.

[13] Yokoyama, Y. et al. 1978. Development and application of multi-layered vibratory bowl feeders. In, *45th Tech. Meeting of the Committee for Automatic Assembly*, pp.31-39 (In Japanese). Japan Society of Precision Engineering, Tokyo.

[14] Yoshida, R. 1983. On the revolving parts feeder with velocity servo-mechanism. In, *Proc. 4th Int. Conf. on Assembly Automation*, pp.209-220. IFS (Publications) Ltd, Bedford, UK.

[15] Kaneiwa, T. 1983. Introduction of automatic assembly line for power steering pump parts. Preprint for, *Tech. Meeting of the Committee for Automatic Assembly*, No. 83-8, 1-7 (In Japanese). Japan Society of Precision Engineering, Tokyo.

[16] Mitsui, N. 1982. Multiple insertion machine of electronic parts. Preprint for, *Tech. Meeting of the Committee for Automatic Assembly*, No.84-8 (In Japanese). Japan Society of Precision Engineering, Tokyo.

[17] Ikeda, M. 1982. Automatic accommodation feeder and its application. Preprint for, *Tech. Meeting of the Committee for Automatic Assembly*, No. 82-3, pp.15-18 (In Japanese). Japan Society of Precision Engineering, Tokyo.

ROBOTIC ASSEMBLY OF WORD PROCESSORS

R. N. Stauffer
Managing Editor of Robotics Today

First presented in *Robotics Today* (October 1982). Reproduced by permission of the author and the Society of Manufacturing Engineers.

The IBM 7565 robotic manufacturing system is proving to be a smart and versatile performer in on-line assembly operations. The robotic assembly system for the assembly of IBM's Displaywriter word processor is described.

Automatic assembly will be a common denominator in most of tomorrow's highly automated factories. The task will be handled routinely by intelligent, computer-controlled robots with tactile and optical sensing features that contribute to flexible, adaptable operation.

The scenario is not reserved strictly for some future setting, however. It exists today in a number of IBM Corporation manufacturing facilities. One is a large plant in Austin, Texas, where the Communication Products Division builds the corporation's Displaywriter word processing system along with other sophisticated information-handling equipment. Three IBM 7565 robotic manufacturing systems are used for on-line assembly in the production of the Displaywriter. A fourth robot, located off-line in a laboratory, is used for building Displaywriter subassemblies and for training and development work.

The robotic assembly installation at Austin is a significant and tangible dividend from a major Automation Research project that IBM has been conducting during the past 10 years. This project is aimed at developing advanced technologies that will be required in factories of the future. Three areas of primary concern include robots, machine vision, and 3-D solid geometric modelling.

The overall effort has borne considerable fruit, particularly with regard to the development of robotic hardware and software. The RS 7565, introduced in 1982, is designed primarily for precision assembly and testing operations. Built by the IBM System Products Division in Boca Raton, Florida, the system is currently being used in a number of IBM manufacturing plants and in several test marketing situations in customer facilities. The following details on its design and operation will be helpful in understanding its application at Austin.

Tactile and optical sensing

The RS 7565 has some innovations that make it uniquely suited for light assembly, drilling, and other operations requiring both speed and precision. Its general configuration consists of a rectangular metal frame, measuring about $1.8 \times 1.2 \times 0.9$m, which houses a hydraulically powered arm. The arm can move a 2.25kg payload at speeds up to (1000mm/s) through six degrees of freedom. A servo-actuated, two-fingered gripper provides a seventh axis.

The systems used at Austin, one of which is shown in Fig. 1, incorporate both tactile and optical sensing, providing the ability to respond to abnormal conditions and continue operating under most of the situations encountered. Strain gauges located in the gripper provide force sensing capabilities at the tip, side and pinch surfaces of each finger. A light-emitting diode (LED) is used to detect the presence of objects between the fingers. It is important to note that the gripper is also used to carry various types of multipurpose tools required in assembly work.

One of the key features contributing to the speed and precision of the RS 7565 is an IBM-designed linear hydraulic actuator. This unit offers the advantages of direct-drive, a high thrust-to-weight ratio, and fast response. The direct drive configuration minimises backlash, contributing to positional accuracy. The self-commutating actuator converts the up-and-down motion of a series of four pistons into a constant thrust in the direction of motion of the robot arm. The addition of a servovalve on this unit turns it into a servomotor with the capability for high speed and repeatability.

IBM's unique robotics language (AML) is used for programming the RS 7565. The system controller is the Corporation's Series 1 minicomputer.

Fig. 1   IBM RS 7565 robot manufacturing system is one of three used for assembly of the Displaywriter

The cart concept

The configuration of the RS 7565 robot lends itself to an unusual and highly efficient approach to assembly operations in a batch-type manufacturing environment such as that found in the Austin plant. Parts, tools, fixtures, and feeders required for a particular job can be mounted on a mobile cart which is then essentially plugged into the robot for accomplishing the assembly task. Various computer programs are called up as required to handle specific jobs.

Volume requirements determine how many self-contained carts are used with a particular robot. One or two carts might be typical where lot sizes are relatively high. It's feasible, however, that some situations could call for a robot to work with 10, 20, or more carts to meet production schedules on a variety of short run parts.

The big advantage in the use of carts is the ability to keep the robot up and running while new setups are being made off-line. The robot is not surrounded with fixed tooling that has to be changed for each job.

Another feature that facilitates changeover from one job to the next is a fast and simple means of registering each cart to data in the computer. This is done with what is called a find post. The robot arm in manoeuvered to bring the gripper into position next to the post. LED sensors in the gripper fingers then detect the edges of the post in three different dimensions corresponding to the x, y and z axes of travel. The procedure is similar to the use of a contact probe on an NC machine tool to establish the location of a workpiece on the table prior to the start of machining operations.

Applications

The Displaywriter consists of five major components – the keyboard, a display, a media box (i.e. the disk drive), a system electronics box (referred to as the SEB), and an optional printer. At this point in time, automated assembly is used in building the media box and the SEB.

Fig. 2 shows the general arrangement of the three robots on the Displaywriter production line. For purposes of this discussion, they are identified as A, B and C. Assembly operations on the media box are performed by robots A and B, operating together in a serial loop arrangement. Robot C works on the SEB. Both components, the media box and SEB, arrive at the robot assembly stations in kit form. The kits are put together manually at other stations and are then routed to the robot assembly area.

Media box kits descend on an elevator and are moved onto a roller conveyor. As indicated, they are routed first to robot A, then robot B, and then back onto the main conveyor for routing to the final assembly line. The SEB kits are routed to robot C only, and from there they go directly to the main conveyor.

The assembly tasks performed by robot A on the media box kit consist of installing eight screws. Vibratory bowl feeders deliver three different types of screws to rotary escapement devices ready for pickup by the robot. Fig. 3 shows the general arrangement of the robot and auxiliary equipment.

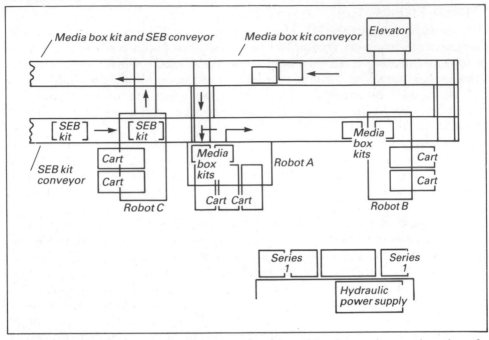

Fig. 2   Layout of the production line – note location of the three robots and routing of SEB kits (to robot C only) and media kits (to robots A and B)

Fig. 3   Robot A installs eight screws in the Displaywriter media box, using a general-purpose multifunction driver. Vibratory bowl feeders deliver three different types of screws to rotary escapement devices

The media box is pulled into the robot's work envelope by an air cylinder. It is then clamped in a fixture, locking the base plate and two disk drives in position ready for assembly. The air cylinder is controlled by the digital output of the robot's Series 1 computer. When clamping is complete, the condition is sensed to provide a digital input back to the computer.

Using a general-purpose, multifunction air driver, the robot removes a magnetic driver bit from its holder and proceeds to drive two large screws to fasten the base plate to the front disk drive in two positions. It then exchanges the larger bit for a smaller unit, drives one smaller screw into the front disk drive and three other small screws into a second disk drive. Next, using the same magnetic bit, the robot picks up two more screws and prestarts them in the base plate. This operation saves time further down the line in a manual assembly area.

The primary task performed by robot B on the media kit is to assemble a plastic strain relief and four rubber feet to the base plate. After the kit arrives at this station from robot A, the base plate is clamped in a fixture. The robot picks up the strain relief (Fig. 4) which was carried loosely in the kit, and places it in the fixture.

Robot B then proceeds to pick up and drive three screws (one at a time) to attach the strain relief to the base plate. In the next sequence, the four feet are acquired (also one at a time) and secured to the base plate (Fig. 5). Screws are fed individually into the feet from a bowl feeder before the feet are picked up by the robot. The last operation performed by robot B on the media box is to start two screws on the bottom of the base plate.

Fig. 4   The plastic strain relief shown in the gripper of robot B will be placed in the base plate of the media box and secured with three screws

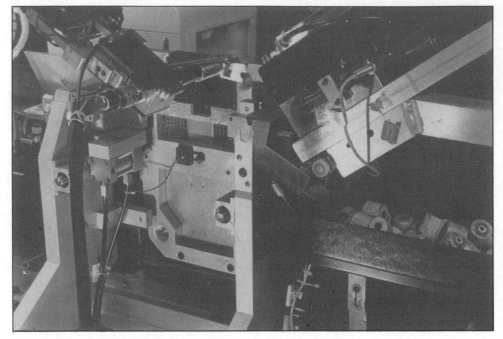

Fig. 5   *View of robot B showing the attachment of one of four rubber feet to the base plate*

Part time work

The versatility of the RS 7565 and significance of IBM's concept of using self-contained carts to carry tools, fixtures and parts, are both evident on robot B. In addition to the operations just described, this robot is also used to assemble a front cover support consisting of three metal brackets and four screws. These components are brought to the robot on carts which are rolled into the robot's work envelope at two positions. The assembly operation is similar to one performed by the off-line laboratory robot described later.

The front cover support subassembly job is run only when the robot is not required to be working on the media kit. It may be used to run this secondary job, for example, during a lunch hour or at night. It is also possible to plug in different carts in order to run subassemblies other than the front cover support. In fact, all three robots on the line have two cart positions which can be used for building other subassemblies in an off-line mode.

Assembling the SEB

The assembly operations performed by robot C on the system's electronics box involve two major assembly tasks. In one, the power supply is assembled to the cover; the other involves the assembly of a motherboard subassembly consisting of a board (used to hold logic cards), two support frames, two metal brackets, and four screws.

The SEB kit comes into the robot work envelope on a conveyor and, along with the box bottom cover, is automatically fixtured by means of a

DO-controlled air cylinder. The robot arm then pushes the power supply into the same fixture, lining up holes in the power supply with screw inserts in the cover.

At this point, the robot gripper picks up a special-purpose, multi-function driver similar to that on robot A. Four screws are picked up and driven in sequence through the power supply and into the bottom cover. During this operation, components for the motherboard assembly are also being fed into position. Specifically, air cylinder-driven pick-off knives are used to pick the bottom board from a stack and place it in a nest.

The robot picks up a card support frame, turns it around, and using the frame itself, picks up two screws from a double-headed screw dispenser. It then moves the frame to the right side of the motherboard. The frame is placed in position and held there with a latch while two screws are driven.

While the first frame is being screwed to the right end of the motherboard, a second frame is fed into position for pickup. The same sequence is followed. The robot carries the frame to the screw dispenser and then on to the other end of the motherboard ready for insertion of two screws.

Fig. 6 shows one of the card support frames being moved to the screw feeder. The robot gripper uses a common driver to assemble both the power supply and the motherboard. The driver is not set aside at any time during the assembly sequence. A special air clamp on the driver allows it to pick up the card support frames as well as drive screws.

Fig. 6 In building a motherboard assembly, robot C picks up this card support frame, moves it to a screw feeder, and screws the frame to a board that will ultimately hold logic cards

Once the second set of screws has been driven, all clamps are released automatically, the driver (via its clamp) grasps one of the support frames, removes the entire assembly from the fixture and places it in the SEB kit on top of the power supply. The kit is then unclamped, and the robot pushes the pallet and kit back onto the main conveyor line.

The ability to sense and respond to abnormal conditions is the key to continuity of production. The robot's ability to recover from such conditions reduces the number of necessary operator interventions. Importantly, this allows the operator to attend a greater number of automatic assembly operations. The robots used in this installation at the Austin plant are programmed to respond to and correct a variety of abnormal situations. Examples of faults detected include screws with missing threads, screws with the slot missing on the head, and holes that are not tapped. Typically, the screw returns with the driver. When the driver attempts to pick up another screw, the robot senses through force feedback that the screw it attempted to drive previously is still in the magnetic chuck. In such a case, an electromagnet removes the first screw, and the driver picks up another screw.

The usual procedure is to try an operation three times. In the event the task is still not completed, it is assumed that a condition exists from which the robot cannot recover. At that point, a signal is given for operator intervention.

One interesting checking routine is used on robot C after the last two screws are driven in the motherboard assembly. The robot is programmed to move the driver back over the lefthand card support frame and down to touch the motherboard with the driver tip to confirm its presence.

Gear plate assembly

As noted previously, a fourth robot is located in the Development Laboratory. This unit is used for running production lots of a gear plate assembly for the Displaywriter. The assembly consists of a metal plate, three plastic gears (two types), and three C-clips to retain the gears on three pins. A technician oversees this installation, coming in several times a day to load parts on the cart. The robot is also used for training purposes and the development work.

The job essentially runs itself. Again, numerous fault routines are included in the program. Faults such as a missing clip or gears that do not mesh initially are corrected automatically. For example, if one of the pins on the baseplate is too short, it may not be possible to install a C-clip. After three attempts, the robot would discard the defective plate and acquire another. And in the situation where two gears do not mesh on the first try, the robot responds by giving the gear in its gripper a slight twist to achieve proper meshing.

IBM engineers, B. Holloway and A. Christy, display understandable pride in commenting on the installation. "Firstly, it's important to remember that we're using robotic assembly on-line as part of production. Obviously, that is a prerequisite for large-scale use of the robot in this role in the future. In our case, we came up with some new concepts in low-volume adaptive assembly, and they helped establish the credibility we needed to get to where we are now. The Displaywriter was not designed for automation. So our approach was to see how much of its processing we could handle with robots without doing any redesign. Although there's still a lot of manual assembly work involved, we have established the basic techniques that would be used with products that are designed for automation. We now have credibility with our product development people. They will have robotic assembly in mind in future product development work."

A lot has been learned in this project. For example, some of the guidelines that might have been given to the product development people to follow for robotic assembly would have created some unnecessary restrictions, i.e. don't use C-clips, or don't use screws, particularly slotted-head screws. But, these and other potential trouble-making features have now been handled successfully and they don't loom as restrictions anymore.

Holloway and Christy point out: "If you do a good job of tooling and programming, then you won't have to spend a lot of time out on the line. IBM engineers devised some effective error recovery routines and did a fine job on the tooling. The real challenge in industry in the next few years will be to find enough people capable of handling the application of robotic assembly in production".

INTEGRATION OF ROBOT OPERATIONS AT FLYMO

M. W. Leete
Flymo Ltd, UK

First presented at the 6th British Robot Association Annual Conference, 16–19 May 1983, Birmingham, UK. Reproduced by permission of the author and the British Robot Association.

Flymo has used robots in its production processes since 1979. The first applications were unloading components from injection moulding machines using ASEA MHUs. During 1981/82 a development project in robot assembly was undertaken with a Unimate PUMA. This was successful and a production assembly cell was constructed employing three robots, two PUMAs for assembly and a Cincinnati T^3 for feeding and palletising. The cell is linked to robot controlled machines making plastic mouldings and tubular steel handles. This paper describes the system and explains that the motivation for the project was not direct labour saving but a desire to reduce overheads by eliminating handling, storage and paper transactions between operations.

The majority of Flymo parts are made from injection moulded thermoplastic, the machines making them operating 24 hours/day. Since 1979 many of these machines have been operated by robots[1]. Traditionally assembly, testing and packing involved 25 operators working day shift only in an assembly shop located in a separate part of the building, scheduling of piece part production and assembly being carried out separately for the two departments. This necessitated a work-in-progress store for component parts and it was normal to have at least one weeks stock. The view was taken that this non-productive and costly stage should be eliminated by integrating assembly with production of piece parts.

To combine the labour-intensive manual assembly with component manufacture was impractical with so many people involved. However integration was possible if the assembly labour content could be reduced by automation. Fig. 1 shows diagramatically how the operation sequence was to be simplified.

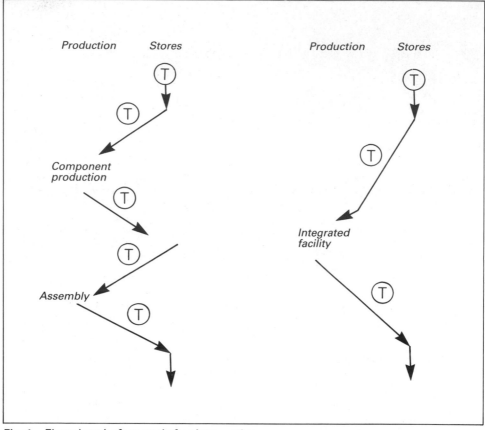

Fig. 1   Flowchart before and after integration

Assembly automation using dedicated machinery was incompatible with the flexibility required in order to cater for future products, therefore a programmable robot system was needed. Such a system was not commercially available so Flymo decided to develop its own using 'off-the-shelf' robotic and computer equipment where possible. With limited experience of assembly automation, detailed design of the fully integrated system was not possible at this time and it was decided to execute the project in three stages:

- *Development* (June 1981/December 1982) – Experiments in robot assembly and trials in production environment.
- *Part integration* (January 1983) – Integration of assembly cell with parts production (major items) and final assembly/test/pack.
- *Full integration* (1983/1984) – Incorporation of facilities for producing minor component parts.

Objectives

The objectives of the exercise were four-fold:

- Reduce direct labour content in assembly operation.

Fig. 2   Parts assembled by PUMA

- Save overheads – 24 hour operation would give higher utilisation of floorspace and other fixed overheads.
- Reduce materials handling – the elimination of storage between component manufacture and final assembly would reduce both indirect labour and fork lift truck requirements.
- Save inventory cost – previous to integration it was normal to hold at least one weeks stock of parts as a buffer between operations.

The cost of holding this stock is calculated as 33% of value per annum. This is made up from: financing value of stock, space cost, and cost of storage containers and racking.

Phase 1 – Development of robot assembly

The objective of phase 1 was to develop a cell to automatically assemble components of the electric motor of the Flymo Minimo (Fig. 2). At this time the details of how the cell would be applied within the final integrated facility were not formulated.

Cycle time

Cycle time would be dictated by the output of the slowest machine producing component parts. This was the moulding of the hood with a cycle time of 1.5 minutes. It meant that for economic reasons only one robot could be employed for the assembly task. As it would have to carry out a number of sequences a servo controlled robot with at least 5 axes was necessary.

Robot selection

The robot chosen was the five axis Unimate PUMA 500 with 8K of battery back-up CMOS memory. It was selected after applying the following criteria:

- Accuracy.
- Physical size – had to be suitable for table mounting.
- Capacity – a minimum of 2kg payload was required.
- Ease of programming.
- I/O capability – communication would be necessary with large number of parts feeder and ancillaries.
- Memory size – it was envisaged that a large number of program steps would be needed during the 1.5 minute cycle.
- Cost – as phase 1 was a development project, strict cash limits had to be applied.

Development period

Although outline schemes had been prepared no details were finalised on the cell until the PUMA was delivered and set up in the development room. Even then little work was carried out on the drawing board, the final working cell being arrived at by progressive development. This work started in June 1981 and took just over 12 months. By the summer of 1982 there was a working assembly cell, although not a particularly efficient one as all parts had to be preloaded, the motor and its mounting bracket by hand, the others through bowl feeders or magazines. On completion the cell was tested in a production environment for six months, serving the purpose of training technical and operating personnel and allowing debugging under working conditions.

Component design

During the development period considerable attention was paid to detailed aspects of component design, fits, lead-ins, etc., and this involved working with design engineers, quality assurance engineers and outside suppliers, as well as internal production departments in order to obtain parts suitable for automatic assembly.

Gripper

After toying with quick-change gripper systems it was found possible to accommodate a wide variety of components within one set of jaws by using a step design. This is illustrated in Fig. 3. Mounting of the gripper to the robot arm was critical; it was found necessary to incorporate X-Y flexibility to take up misalignment of components. The vertical direction incorporates preloaded movement of almost 250cm so that a known force could be applied, and transducers were built into the gripper to ensure that a part had been successfully picked up or assembled. The die cast insert, impeller and motor shaft all had drive flats to be lined up. This was achieved by turning the motor shaft until the parts fitted together.

Fig. 3   PUMA gripper

Driving screws

The mounting bracket is fitted to the motor by four screws. A self feed driver is used and it had been planned that the PUMA would pick this up and drive in the screws but unfortunately the force necessary proved too much for it to

Fig. 4   PUMA with screwdriver and parts feeders

manage. The solution was to mount the driver on an articulated arm with X-Y freedom with the PUMA doing the positioning (Fig.4). Downward force was provided by an MHU linear module mounted between the screwdriver and the articulated arm.

Parts feeders

At the development stage both the motor and mounting bracket were preloaded to a two-position rotary jig. The blade and the impeller were fed from magazines, other parts from vibratory bowl feeders.

Control system

At this stage overall control of the cell was achieved using the PUMA's I/O capability. It was found to be very easy to achieve this with Unimation's VAL language. Although a terminal is not necessary to run the PUMA once programmed it has been the practice at Flymo to leave it connected and to print current status of both robot and ancillaries on the screen. This facility is built into VAL.

Results

By March 1982, halfway through the development period but before the cell was completed, it was clear that robot assembly was perfectly feasible and the decision was taken to proceed with phase 2. This would involve integration of two PUMA assembly cells with both production of components and final assembly/testing/packing.

Phase 2 – Integration of assembly cell with other operations

The principle of the integrated facility is outlined in Fig. 5. There are two PUMA cells carrying out detail assembly of components to the motor. Each operates independently and can produce either the same product or

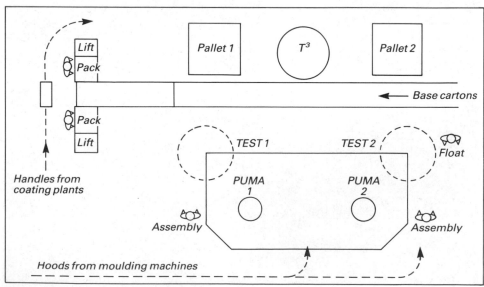

Fig. 5   Plan of integrated facility

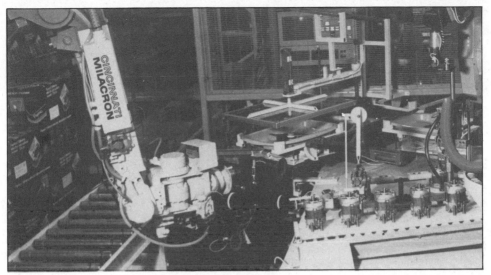

Fig. 6   Cincinnati T³ removes finished product from test, PUMA driving screws

different ones. At present only two different products are catered for but more are possible. The PUMAs are serviced by a Cincinnati T³ robot which also packs and palletises finished products (Fig. 6). A proportion of the work cannot be carried out by the robots and two operators are involved in each build.

The large plastic lawnmower hoods are produced on adjacent robot-operated machines and fed to the assembly facility by roller conveyor. Lawnmower handles are also produced automatically and delivered to the operator at the packing station on a overhead conveyor.

Service robot

It was decided that the robot servicing the PUMA cells should also carry out packing and palletising operations. From the scheme layouts it was clear that the only robot available with sufficient reach was the hydraulic Cincinnati T³ range.

As can be seen from Table 1 the T³ has to carry out a wide variety of tasks. This posed problems in gripper design and control system. However, the table does not show the extent of the problem since at each stage there are branch instructions to cater for any rejects. In addition, the two PUMA cells do not necessarily operate in a fixed sequence relationship with one another.

Gripper

The T³ carries out three different operations on two different products, each necessitating its own pick-up method. First thoughts were to have three different grippers and use a quick-change system. As this could involve up to eight changes in a three minute period the idea was rejected in favour of the multiple gripper arrangement shown in Fig. 7. Mechanical jaws are used to pick up the motor and one of the lawnmower hoods whilst suckers are used

Table 1   Sequences chart

	T3	PUMA	TEST 1	TEST 2	OP 1	OP2
Load motor to table 1	•					
Index table 1 to position 2 and assemble parts to motor		•				
Index table back to position 1			•			
Pass assembled motor power unit to operator 1	•					
Assemble power unit to hood, fit labels and connect cable, place in jig on table 2					•	
Index table 2 to position 2, drive screws		•				
Index table 2 to position 3				•		
Index table 2 to position 3, pack product in carton	•					
Finish pack						•
Palletise	•					

for the other hood and palletising cartons. Although it looks cumbersome the T^3 wrist arrangement allows easy selection of the gripper segment required.

Supervisory control system

The problem was communication between the five computers within the facility (two for PUMA, one for the T^3, and two electrical testers). It would be possible for them to communicate direct with each other in a matrix but this would not be easy as the overall sequence is not fixed and trouble-shooting control system problems would be difficult.

Fig. 7   Cincinnati T^3 gripper showing three pick-up mechanisms (two mechanical, one suction)

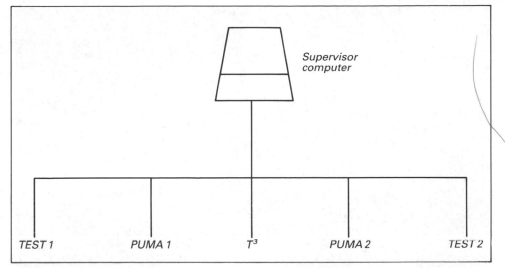

Fig. 8   Computer hierarchy

The solution finally adopted was the provision of a supervisory computer operating at a higher level than the control computers. This hierarchy of control is shown in Fig. 8.

The role of the supervisor is to keep track of operation status and call each device at the correct time, notifying it which routine it has to execute. In addition to the sequences shown in the chart there are variations required if a unit fails a test or the PUMA detects it is unable to assemble. The supervisory computer displays current system status and provides simple error diagnosis and production records.

The requirements for the supervisory computer are as follows:

- Ease of programming by non-specialist personnel.
- High-level language capable of supporting VDU display and communication with peripherals such as floppy disk and printer for software development and data logging.
- Facility to develop software off-line.
- Potential to be incorporated in local area network.
- Reliability.

It was felt that the normal PLC was unable to meet this specification. Consideration was given to using an industrial specification single board computer but previous experience with such systems was unfavourable. A high degree of specialist technical knowledge and an expensive development system is required to obtain the best out of such equipment; neither was available in the plant and support from manufacturers was found wanting.

Flymo has been using general purpose microcomputers for some time in data logging applications. These had been successful and it was decided to use such a machine, a Commodore 4032 (Fig. 9), as the supervisor for the new facility. Similar machines were also to be used for controlling the electrical testers.

Fig. 9   Modified Commodore 4032, I/O board at front, Rompager at rear

Computer interface hardware

The Commodore 4032 can be connected to the outside world in a variety of ways. In this application an interface board custom designed by Flymo was used providing 16 digital I/O channels and two analogue inputs. The latter were not required for the supervisory computer but were necessary for the electrical testers and applications elsewhere. The board was connected to

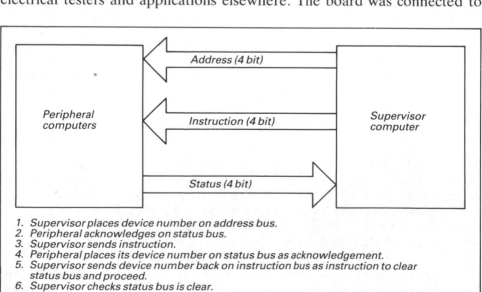

1. Supervisor places device number on address bus.
2. Peripheral acknowledges on status bus.
3. Supervisor sends instruction.
4. Peripheral places its device number on status bus as acknowledgement.
5. Supervisor sends device number back on instruction bus as instruction to clear status bus and proceed.
6. Supervisor checks status bus is clear.

Fig. 10   Bus structure and handshake routine

the computer via one of the spare expansion ROM sockets. Optical isolation is provided for inputs and 24 volt drivers for outputs.

The computers communicate with each other through the bus structure shown in Fig. 10; this operates at 24 volts. The PUMA and the T^3 are connected via their I/O connections.

Computer software

The software is written in BASIC and this has been found fast enough for the sequential control required. Once developed the program is stored in EPROMs mounted on a JCL ROMPAGER board which boots the program automatically into RAM on power up or reset.

BASIC is not an efficient language and discipline is required during programming to create a structure that is understandable and therefore easy to modify and maintain. Its advantage lies in the increasingly large number of people who, through the various computer literacy programmes, are able to work with the language. If speed becomes a problem then the program can be compiled or sub-routines written in machine code, although the latter approach requires considerable experience.

Phase 3 – The future

Further integration

The benefits of integration are such that the principle is being extended to incorporate some of the smaller parts. To do this it is necessary to balance cycles by employing quick change mould systems or using family tools.

Further automation

The robots will take on additional work within the facility. The first job was mating the hood to power module automatically. This allows the manual operations to be grouped at the completion of the robot operations, apart from palletising.

Flexibility

The cell presently contains a significant proportion of dedicated equipment which restricts production capability to two models. Work is being done to reduce dedicated equipment and increase flexibility.

Concluding remarks

Flymo sees the achievement thus:

- Integration of robot assembly process with production of components reduces non-productive operations and therefore overheads.
- Reduced assembly labour content allowing assembly over three shifts.
- Successful application of general purpose robots to complex assembly tasks.
- Simple gripper system on assembly robot incorporating non-rigid mounting and sensing.
- Use of general purpose 'personal' computer for supervising robots and controlling product testing.

The project has not been an easy one and it cannot be regarded as completed because it can still be improved. The implication is therefore that it will never be complete and that must be, for in competitive business there are always improvements to be made and technology does not stand still.

References

[1] Leete, M. W. 1981. The Electrolux MHU in injection moulding – A case study. In, *Proc. 5th British Robot Association Annual Conference*, May 1981. British Robot Association, Bedford, UK.

ROBOTIC ASSEMBLY OF KEYCAPS TO KEYBOARD ARRAYS

G. J. VanderBrug and D. Wilt
Automatix Inc., USA
and
J. Davis
Digital Equipment Corporation, USA

First presented at the 13th International Symposium on Industrial Robots and Robots 7, 17-21 April 1983, Chicago (SME Technical Paper MS83-328). Reproduced by permission of the authors and the Society of Manufacturing Engineers.

A two-station robotic assembly system for inserting keycaps into keyboard arrays has been designed and constructed. The system is programmable in that it is capable of assembling different styles of keycaps and different layouts of the keyboard arrays. A vision system is used to verify the correctness of the keycaps, and to inspect the keycaps for major visual defects.

Assembly is one of the most challenging of all robotic applications, because of the diversity and complexity of the processes it encompasses. Assembly involves more than the insertion or mating of parts – it includes material handling, parts orientation and presentation, inspection, material tracking, and man-machine interaction. Often, these related processes are more critical than the parts mating and insertion operations.

This diversity and complexity has tended to limit the use of automated assembly to high volume applications which have the payback necessary to finance the custom engineering and machinery required. Now robotic assembly, however, is reducing the volumes required to justify automation by replacing fixed automation cams and actuators with the soft cams afforded by servo controlled motions. Furthermore, the development of robot vision has provided a sense more powerful than any hard automation system and which can be fully utilised only by the flexible motion capabilities of a robot. Nevertheless, this flexibility comes at the expense of speed, since robotic assembly systems typically run more slowly than hard automation assembly machines.

This paper describes a robotic system for the assembly of keycaps onto keyboards. The components of the system involve all of the previously mentioned aspects of assembly. It offers flexibility in assembling a variety of keyboard models. The system employs the Cybervision[1] assembly system, and was built by Automatix in cooperation with Advanced Manufacturing Engineering, Digital Equipment Corporation.

The problem involves assembling keycaps onto keyboards. The keycaps come in a variety of legends, colours and shapes. They are procured from several vendors, a factor which introduces both subtle and obvious variations in the keycaps. The keycaps are shipped in bulk according to legend, colour and shape. However, there is no assurance that a container contains only a single legend, nor is there any guarantee of the quality of every keycap. The system must work with a variety of keyboard styles, which differ both in the keycaps placed in given locations and in the geometry of the locations.

The system

In order to assemble keycaps onto keyboards, it is necessary to acquire the keycaps, then orientate, inspect, and insert them in the keyboard. These tasks were divided between two separate stations. The first station orientates keycaps, inspects them, and loads them into magazines. Since vision plays an integral role in these tasks, it is called the vision station. The vision station involves the material handling, parts presentation, inspection, material tracking and man-machine interaction aspects of assembly.

At the second station, keycaps from the magazines are inserted into keyboards by the robot. This is called the robot station. The robot station involves material handling, parts mating, material tracking and man-machine interaction aspects of assembly.

The interface between the two stations is done with carts and magnetic tapes. A single cart is used to transport a set of magazines that consititute a

Fig. 1   An overall view of the keyboard assembly system. The vision station is on the left; the robot station on the right; the cart is in the left foreground

Fig. 2 The vision station, showing the Autovision II processor and three of the four magazine indexing units

single job from the vision station to the robot station. A magnetic tape, containing information about the keycap styles and the number of keycaps in the magazines, is associated with each cart.

An overall view of the Cybervision assembly system is shown in Fig. 1. The vision station is shown on the left; the robot station on the right; a cart is shown in the left foreground. The vision station is comprised of the Autovision II system, four bowl feeders, four cameras, and four devices for locating and indexing magazines. The robot station consists of the AI32 controller, the AID600 robot, end-effector tooling, a parts presentation system and a dual shuttle for transporting keyboards. These components are shown in Figs. 2 and 3.

During the initial stages of this project, several system architectures which

Fig. 3 The robot station, showing the AI32 controller, the presentation array, and the AID600 assembly robot

integrated the parts presentation and inspection with the parts mating into a single station were considered. However, the system selected establishes a buffer of orientated and inspected parts in the magazines that are stored in the carts. This buffer enables the individual stations to be run at their maximum rates, and makes them more independent of each other with respect to downtime. In addition, this two-station approach makes it easier to add on to individual components of the system. Also, it allows for the possibility of obtaining magazines of orientated and inspected keycaps from alternative sources in the future.

Both the vision station and the robot station use a sequence of menus which appear on the CRT and help to guide the operator through the proper operation of the systems. The menus involve the normal operation of the systems, the setting up for new keyboards (including training for inspection on the vision station and training for which keycaps go in what locations on the robot station), certain system utilities, and functions for changing parameters.

Both stations maintain data on magnetic tape which describes how to perform each task for a particular style of keyboard. For example, the vision station must be told which keycap to place in each magazine, which bowl feeder to use for inspecting the keycap and what to look for when inspecting the keycap. The robot station, on the other hand, must know where to pick up keycaps, which fingers of the end-effector to use, and where to place the keycaps on the keyboard. Each system maintains this data in a group of files called a JOB. For each style of keyboard there is a JOB for the vision station and a JOB for the robot station.

Since there are many keyboard styles which can be assembled, the system has been equipped to handle many JOBS. A single file called the JOB DIRECTORY keeps track of the number of JOBS that exist and what file names are used by each JOB in storing the data. When the systems want to find data about a particular keyboard, they first use the JOB DIRECTORY to find the file names that are used by the JOB. Once the file names have been found, the system reads the JOB data from those files.

The vision station

Components

The heart of the vision station is the Autovision II processor. It provides computer processing and RAIL programming for performing the visual inspection task[2] (see also page 219), for maintaining data bases for the different keyboard styles, for controlling other components of the system, and for providing for the menu-driven operator interface. There are four bowl feeders, one for each keycap shape (this being a function of the keyboard row). Four cameras are used to verify the correctness of the keycap legends and to inspect the keycaps as they emerge from the bowl feeders. Incorrect or defective keycaps are diverted to a reject bin while good keycaps are sent to the magazines. Each magazine has eight rows of keycaps and can contain up to 250 keycaps. Up to 80 magazines can be used for a single job. Fig. 4 is a close up of one of the four cameras and its inspection track.

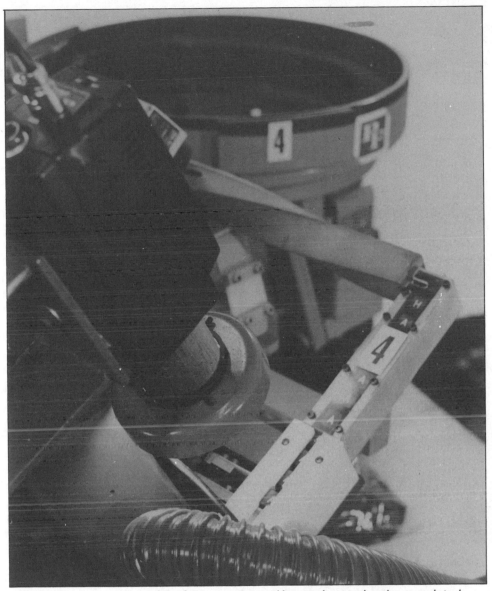

Fig. 4   A close up of one of the four cameras and inspection tracks; the associated bowl feeder is in the background; rejected keycaps are blown out the tube in the foreground

Normal operation

To build a batch of keyboards, the user inputs the keyboard style and the number of boards to be built. The system then uses the JOB DIRECTORY to access the proper data base for that style of keyboard. From this point on, the system instructs the user (via the CRT) regarding the proper placement of keycaps into the bowls and the placement of the magazines into the indexing mechanisms. (Each magazine set is labelled from 1 to 80.) A sample print-out of the CRT is shown overleaf.

Bowl Feeder	#	Current Legend	Mag	#	Next Legend	Mag	Remove Mag	Status
1	45	Break	13	44	Bkspc	14		Active
2							30	Done
3	55	> •	26	71	Noscr	27	25	Active
4	01	A	34	51	＼ ?	35	33	Service

Each bowl feeder has a CURRENT and a NEXT column. These columns show the keycap part number (in instances where they provide sufficient identification, only the last two digits are shown), the keycap legend, and the magazine number. The CURRENT column contains information about what is currently active at the respective bowl; the NEXT column contains information about what is to be loaded next. When a magazine is completed, it is removed from the CURRENT column, and its number is placed in the REMOVE column. Normally a bowl will show an ACTIVE status, indicating it is in the process of loading keycaps into magazines. Once the magazines loaded by a bowl feeder have been filled, the STATUS will be set to DONE. If, during normal operation, the vision system detects a status that requires the operator's attention (such as a bowl feeder which has stopped feeding parts, the occurence of a jam at the indexer, or a magazine has been filled with the proper number of keycaps), then the word SERVICE appears in the status column and a light on top of that bowl feeder station goes on.

It is worth noting that a single bowl feeder is used for orientating many distinct keycaps (all of those that have the same shape). When it is time to load a different keycap into a bowl, the operator removes any current keycaps from the bowl before loading the new ones. However, if all of the previous keycaps are accidently not removed from the bowl, they will be rejected by the vision system for having the incorrect legend.

It takes less than 10 minutes for a magazine to be fully loaded. Since there are four bowl feeders and cameras, the vision station is capable of orientating, inspecting, and counting over 1000 keycaps in 10 minutes. The actual time varies according to the keycap style (some legends take longer to inspect than others). In addition to counting the good keycaps the system also keeps statistics on the bad keycaps.

Training on new keycaps

To train the system to recognise new keycaps the operator selects the KEYCAP TRAINING MENU. This leads him through a sequence of steps which determine certain parameters, such as: is the keycap a light legend on a dark background or vice-versa; the proper window which contains the legend; certain threshold settings; and the features to be used in the inspection process. Once these parameters have been established, the operator loads a small set of good keycaps in the inspection track. The system will process these good keys and set up the proper data base. Fig. 5 shows three examples of keycaps that are rejected by the system.

Fig. 5 *A close up of three keys that represent those rejected by the vision system. The defect in the first keycap is in the 'B', the second keycap is defective at the upper left of the legend, and the third has a smear across the '1'*

The robot station

Components

The robot station consists of the following components: AI32 controller, AID 600 robot, end-effector tooling for handling many keycaps at one time, dual keyboard shuttle systems, keycap presentation array, and remote operator control station (see Fig. 3).

The heart of the robot station is the AI32 controller. It provides the computer processing and RAIL programming used in controlling the robot, the end-effector tooling and other components of the system, in maintaining the keyboard data bases, and in providing the menu-driven operator interface.

The AID 600 robot is configured for this system as a three axis (x,y,z) robot. Its Cartesian kinematics, high repeatability and accuracy, and large tool load capacity, are well suited for this application.

The end-effector is a flexible tool for picking up keycaps from the presentation array and inserting them in the keyboard. In general, the more keycaps that are inserted at a time, the faster and less flexible the system becomes. Inserting an entire row at a time is faster and less flexible than inserting a single keycap at a time. The chosen design keeps some of the best of both worlds in that it picks many keycaps from the presentation array, but inserts a smaller number (from one to four) at a time in the keyboard. It reduces the number of the long, time-consuming moves to the pick-up station, but maintains flexibility of where these keycaps are inserted in the

Fig. 6   *The part presentation array and the dual shuttle system*

keyboard. Individual fingers in the end-effector are under software control. Vacuum is used to hold the keycaps. The end-effector design results in the assembly of one keycap in less than a second, but maintains system flexibility over a large variety of keyboards.

The two shuttles provide a safe and convenient method for presenting blank keyboards to the robot and for receiving assembled keyboards from the robot. Clamps and pins hold the keyboard in place on the shuttle fixture. The shuttle can be configured for different keyboards (when necessary) by changing the fixture on which the keyboard is placed. The use of two shuttles allows for continuous assembly operation by the robot. The shuttle system and the presentation array are shown in Fig. 6.

The presentation array is an 80×8 matrix of keycap pick-up points. Each column is used for the same keycap, although they can be adapted to allow for fewer quantities of a larger number of keycap styles. There is storage for more than one magazine load of keycaps at the presentation array. This allows for approximately one half shift of operation before loading additional keycaps. When the supply of a given keycap begins to run low, the system warns the operator. A new magazine can be loaded into the presentation array while the robot is assembling.

The operator interacts with the system at the AI32 and at a remote operators panel. Interaction at the AI32 is used for starting up and shutting down the system, initialising for a new batch, and training for a new style keyboard. The nature of the interaction at the AI32 is similar to that of the vision station and will not be described in detail. The operators station includes a foot switch (to inform the system that a blank keyboard has been loaded on the shuttle), a stop switch (to stop the assembly process), and emergency stop switch, and certain status lights (including one which indicates that the AI32 has displayed a message that the operator should read).

The shuttle system is set up so that, at most, one shuttle is at the operator station at any time. This design eliminates any hazard of pinch points at this station. Pressure sensitive safety mats are placed on the floor behind the robot. An additional level of safety is provided by a system of infrared light beams that form an optical fence.

Normal operation

During normal operation of the robot station, the data bases for the particular keyboard to be built are first loaded into the system. Keycaps are then loaded from magazines into the presentation array and data is read into the AI32 from the accompanying tapes. Keyboard arrays are manually placed on the shuttle fixtures and transported to the robot at the command of the operator and the AI32. (The operator depresses the footswitch which signals that there is a key array ready to be assembled, the controller then actuates the shuttle.) When it senses that a key array is in position, the robot assembles the keycaps, and returns the completed keyboard to the operator (provided that the other shuttle has been sent to the robot). The operator installs odd shaped keys, then places the completed unit aside and loads a new array. This process continues until more keys are required, at which time the controller signals the operator, who notifies the person responsible for loading keycaps. The operator then continues to build keyboards while the keycaps are being loaded. If the keycaps being loaded are for the same job, the assembly process will simply continue. If a new job has to be started, the robot will stop at the end of the current job and signal the operator. At this time, information on the new job (containing data on the size, the number of assemblies, the key locations in the presentation array and the location where they are to be inserted in the key array) is loaded and the job is started as before.

References

[1] VanderBrug, G.J. 1982. Cybervision: A system for flexible robotic assembly. In, *Electro/82*. Electronic Conventions, El Segundo, CA, USA.
[2] Franklin, J.W. and VanderBrug, G.J. 1982. Programming vision and robotic systems with RAIL. In, *Robots 6 Conf.*, pp.392-406. Society of Manufacturing Engineers, Dearborn, MI, USA.
[3] Reinhold, A. and VanderBrug, G.J. 1980. Robot vision for industry: The Autovision system. *Robotics Age* (Fall 1980): 22-28.

APPLICATIONS OF ROBOTS TO ASSEMBLY LINES

T. Niinomi and T. Matsui
PERL (Hitachi Ltd), Japan

First presented in *Hitachi Review* (Vol.32, No.5, 1983). Reproduced by permission of the authors and Hitachi.

Rapid progress has been made recently in the automation of assembly work using industrial robots to meet the rising needs of industry for flexible, automated production systems. However, assembly work involves a more diverse set of tasks than other types of industrial work. This has the effect of placing additional economic and technical constraints on progress, in effect holding back the pace of robotisation, and slowing the move towards factory automation. Hitachi, too, is working to develop robot systems geared to specific types of production. In this paper, four such systems are discussed. In robotising assembly work, either of two general types of robot system are adopted: for mass production, systems consisting of a large number of simple, high-speed robots performing highly specialised tasks; and for low- and medium-quantity production, assembly centres using versatile robots provided with automatic hand change, sensor control, and other functions.

In Japan, industrial robots were first used in spot-welding work on automobile bodies. However, with the growing recognition of their vast potential in industry, robots have been adopted for use in materials handling in pressing and moulding work, painting, and arc welding. Recently, robots have even taken on the lead role in plans for factory automation.

However, in spite of their general acceptance in industry, robots are not yet fully used in assembly work. The robotisation of assembly remains an important area for further technical progress.

Assembly work and robotisation

One big difference between assembly work and other types of industrial work that has a direct bearing on progress in automation using robots is the precision handling of parts.

Robotisation has been successfully carried out only in types of work such as welding and painting, in which all that is required is control of the tool trajectory; that is, work that either involves no handling of or contact with the workpiece or at most, handling work such as materials handling, which does not require precise positioning.

In assembly work, however, because parts must be joined together, such features as precise handling and absorption of the reaction forces that arise when coming into contact with the workpiece are required in a robot system. However, the shapes, dimensional tolerances, materials, and precision of fit of the parts to be assembled vary from piece to piece, necessitating diverse, flexible functions in the robot or robot hand. This has resulted in technical and financial constraints that have often made the robotisation of assembly work difficult.

However, the need for robotising assembly work has grown steadily year after year for two main reasons:

- There remain assembly tasks that cannot be automated with current technology in spite of the many manhours required to perform them; these tasks serve as bottlenecks to integrated automation.
- Manufacturing systems making use of robots must be set up in order to meet the growing needs for flexible automated production facilities.

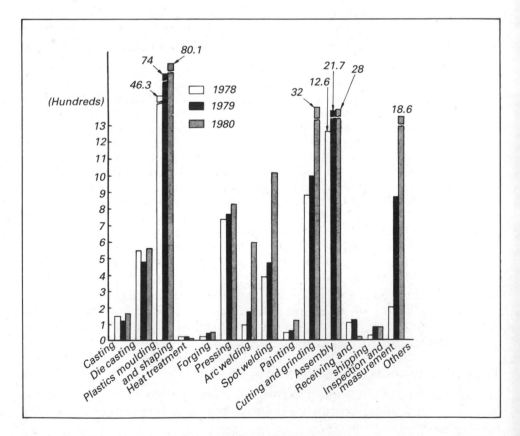

Fig. 1   Number of robots delivered by year and application – the number of assembly robots is second only to materials handling robots used in general industrial work

It is for reasons such as these that many companies are experimenting with robot systems capable of dealing with individual variations in the shape and other features of the products handled, and are bringing in large numbers of simple robots for assembly processes in mass production systems that lend themselves well to line organisation. This is clear also from Fig. 1,[1] in which the number of robots used for assembly work shows a year-to-year increase from 1978 to 1980. Most of these robots, however, are of the sequence type used as specialised equipment; only a few are of the playback type. Robotisation today is failing to keep pace with the demand for versatile robots in batch production with small-sized batches.

In addition to the development of the robot itself[2,3,4] the authors are also engaged in a variety of robotisation projects, including the application of robots to the assembly of mass-produced and individually produced articles. Several representative examples of robot systems developed at Hitachi are described.

Examples of robot applications to assembly lines

VTR mechanisms (mass production)

As is shown in Fig 2, a video tape recorder (VTR) mechanism is made up of pressed parts such as a chassis and a lever, electrical components such as a motor and cylinder head, and also flexible parts such as plastic parts and rubber belts. Prior to the introduction of robots, assembly producibility of the product was evaluated by a method developed at Hitachi[4], and a number of improvements carried out, including modifications to give a construction with higher evaluation indices; reduction in the number of parts

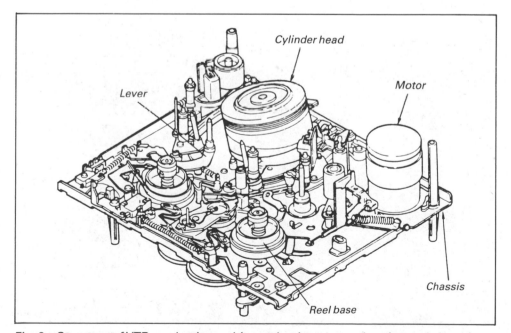

Fig. 2   Structure of VTR mechanism – this mechanism uses a chassis consisting of plastic 'outserted' (moulded) through a steel plate

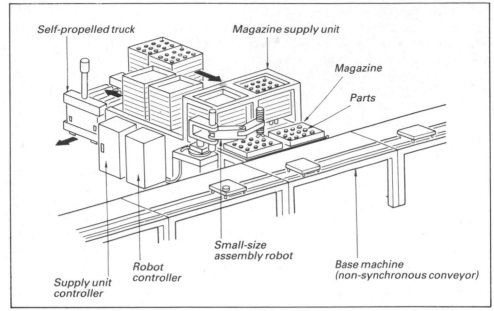

Fig.3   Automated VTR mechanism assembly line – the U-shaped assembly line has a total length of 150 metres; the base machines use a non-synchronous direct feed system which does not use platens

by combining part functions; redesign of subassemblies by the functional unitisation of complex parts; and interconnected electrical components. These changes simplified assembly work, making possible automated assembly in short cycle times.

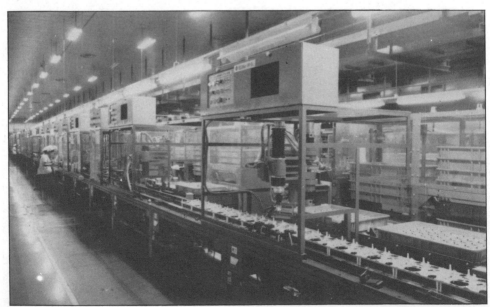

Fig. 4   External view of robot station – this assembly line uses 11 small-size assembly robots

Table 1   Assembly robot specifications*

Item	Specification
Robot	
structure	3-DOF articulated
maximum velocity	1500mm/s
repeatability	±0.05mm
load capacity	20N
Controller	
programming method	teaching playback
position control	software servo
interpolation	linear interpolation

*This robot is economical and rapid working

The automated system, shown in Fig. 3, consists of workpiece transport and positioning base machines, parts supply units, robots, and other automated assembly units. The base machines are designed so that they can stand alone as assembly stations; a non-synchronous assembly line of the desired length can be set up by connecting any number of these in tandem. Parts supply is carried out by vibratory bowl feeders, or in the case of partially assembled units and electrical components, in orderly rows, on magazines, and transferring these on self-propelled trucks from the warehouse to the assembly station.

The automated assembly units consist of a screw fastening machine, an oil applying machine, a pick-and-place unit, and small-size assembly robots. The robot station is made up of a base machine, a robot, and a magazine supply unit. The magazine supply unit places a magazine in a given position, then the robot grasps one part from the end of the magazine and fits in into a chassis positioned on top of the base machine.

Fig. 4 is an external view of the robot station. For economical mass production assembly lines, a small assembly robot having three degrees of

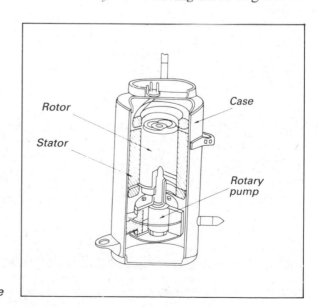

Fig. 5   Structure of a T-type compressor – a rotary compressor used in room air conditioners, it consists of a rotary pump, a motor, and a case

freedom (developed by Hitachi) was used (Table 1). This makes it possible to improve the flexibilty and spatial efficiency required for product changes on the line.

Machine components (medium-volume production)

The product made on this assembly line is a T-type rotary compressor used in room air conditioners. This compressor consists of a rotary pump, a motor, and a case (Fig. 5). Rotary pump assembly tasks include insertion, positioning, centering, screw fastening, press fitting, and shrink fitting. The parts vary in shape from circular to rectangular in cross-section, and most of the parts require precise insertion. Thus, on this automated line, it was decided that such processes as screw fastening, press fitting, and shrink fitting would be performed by peripheral machines, and that travelling assembly robots would perform precise insertion, the loading and unloading of parts to and from the peripheral machines, and transport between the different assembly stations.

Fig. 6 [2] shows the rotary pump assembly line, and Table 2 gives the specification for the travelling assembly robot. This system consists of three travelling assembly robots, a circular track, peripheral machine for screw fastening, etc., and a host microprocessor. While the travelling assembly robot makes one circuit of the continuous track, it exchanges parts with the

Fig. 6   Rotary pump assembly line – this line consists of three travelling assembly robots, peripheral machines, and a host microprocessor

Table 2   Travelling assembly robot specifications*

Item	Specification
Robot	
structure	5-DOF articulated (exc. travel)
maximum velocity	800mm/s
repeatability	±0.3mm
load capacity	70N
travelling velocity	1000mm/s
weight	150kg (inc. travelling base)
Controller	
programming method	teaching playback
CPU	HMCS 6800
position control	software servo
interpolation	linear interpolation, trajectory smoothing
weight	50kg

*The robot and controller are combined into one integrated unit that travels over a circular track

peripheral machines and completes the assembly. The features of the units making up this system are:

- An articulated, suspended-type robot having five degrees of freedom was used. This was given a large operating space.
- A straddling-type travelling mechanism riding on two rails forming a track was employed to move the robot. This was given a bogie truck construction for smooth travel over the curved track. Hard urethane wheels were used to prevent noise and vibrations.
- To reduce the controller size and weight so that it could be coupled to and travel together with the robot, a pulse width modulation servo-amplifier using metal oxide semiconductor field effect transistors (MOSFET) was used as the controller. Also, for trajectory smoothing and to increase the velocity of robot motion, parabolic interpolation control was developed and provided together with the point-to-point control used in conventional assembly robots.
- A single, multipurpose hand was developed that can grasp a variety of parts. This hand has three grippers that can be used selectively depending on the shape of the part being handled.
- Because precise insertion to a clearance of 12–28μm is difficult to achieve

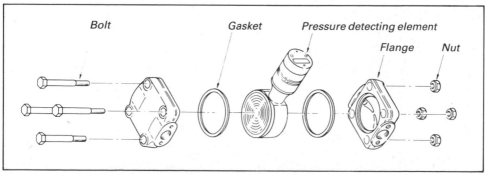

Fig. 7   Flange assembly of differential pressure transmitter – although this unit is made of only 11 parts, there are 4 types of parts, and 5 types of products

in the robot proper, a close-fitting table that enables proper fits even when the fitting point changes was developed.
- An optical transceiver using LEDs and hot transistors was developed to permit data exchange between the robot and the host microprocessor at all points.

Instruments (small-lot production)

The product assembled on this line is a differential pressure transmitter used as the main detector for pressure, flow rate, and liquid level in plant instrumentation. To meet diverse customer needs, this transmitter requires the broad application of advanced design and production technology, in addition to which high reliability is crucial. In automating the assembly of this transmitter, the flange assembly process was selected, which requires high-torque, uniform screw fastening to obtain uniform airtightness, and developed a flexible automated assembly system using a robot. As shown in Fig. 7, flange assembly consists of sandwiching the pressure detecting element between two flanges, each fitted with a gasket, and uniformly fastening these with bolts and nuts. An assembly centre system with the robot at the centre was adopted to permit a retooling-less automation for batch production with small batch-sized and various kinds of products.

Fig. 8 shows a schematic view of the automated assembly system. This system consists of an assembly robot, an automatic screw fastening machine, and parts supply and discharge units. The assembly robot grasps parts

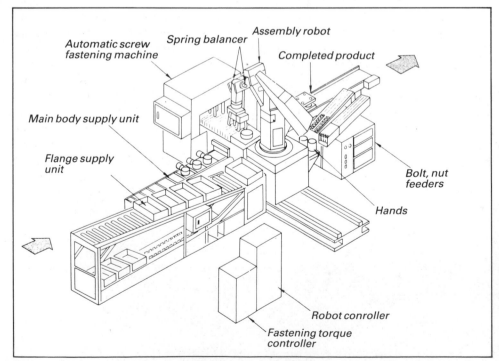

Fig. 8 *Automated flange assembly system – the robot is located at the centre, with parts feeders, screw-fastening machine, and discharge unit located around it*

Fig. 9 Automated flange assembly equipment at work — the robot is unloading a flange assembly that has been bolt-tightened

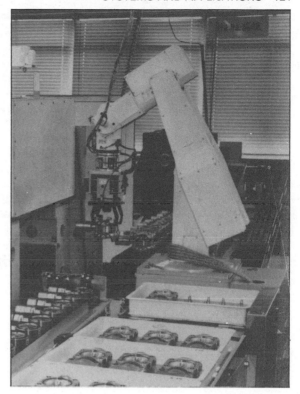

supplied by the four parts supply units in succession, attaches them to an assembly jig, then tightens the bolts with a multi-axial automatic screw-fastening machine, and discharges the finished product. The assembly equipment is shown at work in Fig. 9.

A small-sized articulated assembly robot with six degrees of freedom developed at Hitachi was used in this process (Table 3). A 6-DOF robot was used because of the small size of the parts, the lack of work space, and the complexity of the parts handling. The weight of the parts is a maximum of about 3kg, but because the weight of the flange assembly is about 10kg, the wrist was hoisted up using a spring balancer to reduce the load on the motor.

Table 3 Small-size assembly robot specifications

Item	Specification
Robot	
structure	6-DOF articulated
maximum velocity	1200mm/s
repeatability	±0.1mm
load capacity	30N (150N with use of a spring balancer)
Controller	
programming method	teaching playback (robot language)
position control	software servo
interpolation	linear interpolation

This increased the weight transferrable by the robot to 15kg. Because the robot handles various kinds of parts, it was provided with two hands, one for handling bolts and nuts, and the other for handling the flanges and pressure detecting element. A double-handed automatic hand change system was adopted in which the flange-pressure detecting element hand is attached to the robot arm; when necessary, this hand grips and uses the bolt-nut hand. An automatic self-centering function was added to the end of the robot arm to correct for the position error of parts during assembly, increasing the flexibility of the arm.

Heavy parts assembly line

The traction machine assembled on this line is the driving unit (hoisting machine) used to raise and lower the passenger car of an elevator. This unit consists of a motor, an electromagnetic brake, a worm reduction gear, and other parts (Fig. 10). This is a heavy assembly that is produced in small numbers, and consists of parts ranging in weight from washers weighing just several grams to a 290kg motor. Assembly is carried out by first assembling the worm shaft and sheave shaft on a subassembly line. These subassemblies are then attached at their respective stations to the machine body on a main conveyor, following which the sheave, back drum, and other parts are added. This assembly line is a mixed model line that produces both small and medium-size traction machines. As a result, the development of a low-cost, flexible assembly system was required in automating this assembly process.

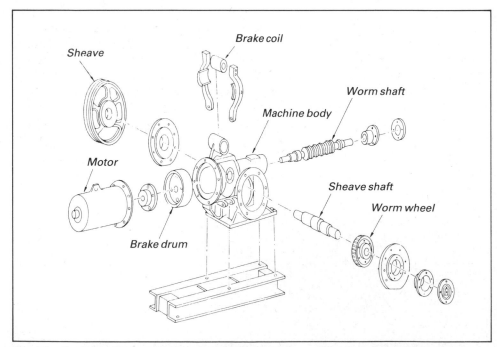

Fig. 10   Exploded view of elevator traction machine – this assembly is made up of a motor, electromagnetic brake, worm reduction gear, and other parts; the automated assembly unit developed is being used in the assembly of the worm shaft

Fig. 11 Automated assembly system for elevator traction machine – this assembly equipment, consisting of a balancer-assisted robot and peripheral equipment, was developed as an automatic assembly system for products that include both light and heavy parts

As shown in Fig. 11 [3], the heavy parts assembly system consists of a balancer-assisted robot and peripheral equipment for bearing insertion and nut-tightening. Of the traction machine assembly processes just described,

Table 4   Balancer-assisted robot specifications*

Item	Specification
Robot	
structure	5-DOF articulated (process robot)
load capacity	100N
repeatability	±0.2mm
maximum velocity	100mm/s(200mm/s when cooperation)
Balancer	
structure	parallel link-type balancer
load capacity	1500N
Controller	
programming method	teaching playback (robot language)
CPU	HMCS6800
position control	software servo
interpolation	linear interpolation, trajectory smoothing
Others	
hands	three hands
automatic hand change	ball lock mechanism

*This robot handles parts exceeding the load capacity (100N) of the small robot by having the balancer bear the weight

this assembly system is used for the partial assembly and insertion of the worm shaft into the machine body.

The special features of this assembly system that makes use of a balancer-assisted robot include the following:

- Light parts are assembled on the body by the robot alone, while heavy parts that exceed the load capacity of the robot are handled by the robot and balancer acting in cooperation with each other; the balancer bears the weight of the part, greatly increasing the transferrable weight of the robot.
- The assembly of various parts is performed by a single robot through the automatic changing of hands specialised for light and heavy parts.
- The balancer-assisted robot can be mounted on a mobile truck, expanding its range of motion.

Table 4 gives the main specifications for the balancer-assisted robot.

Cooperation between the robot and balancer is carried out as shown in Fig. 12. A force sensor is attached to a spring plate installed in the balancer; the sensor consists of four strain gauges. When the robot moves while holding one end of the sensor, the sensor detects the positional deviation that arises between the robot and the balancer. In this way, the vertical direction of drive and velocity of the electrically powered balancer is controlled, and the balancer made to follow the motion of the robot. The strain gauges detect the amount of strain that arises between the robot and the balancer, thus enabling control of the velocity of the electrically powered balancer.

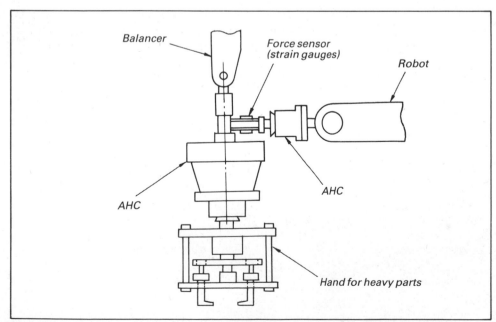

Fig. 12   *Robot and balancer cooperation detector – a force sensor held by the robot is attached to a spring plate installed to the balancer; when the robot moves while holding one end of the sensor, the strain gauges*

Topics for further investigation

The most common requirement for further progress in the robotisation of assembly processes is for improved economy. Next is the ability for highly precise assembly, the ability to keep up with the assembly line speed, and the ability to assemble while making adjustments. There is thus a strong demand for greater cost-effectiveness and performance in robots, and a concomitant need, in the interest of the skillful use and management of robots, for the fuller development of robot software, such as robot languages that are easy to use and have common specifications.

Other areas that require attention include the development of peripheral equipment such as parts supply units, hands, and automatic hand change units, efficient task division planning for robots and dedicated machines, and also parts design as well as assembly construction that is conducive to automated assembly.

Last of all, because of the possibility of work-robot and equipment-robot interference during assembly processes, the development of safety technology in robots is essential.

Concluding remarks

Having discussed the applications of robots to assembly lines by describing several representative examples for different types of production lines, the main points may be summarised as follows:

- The application of robots to assembly lines promises to continue to be an important area of technical development in the future, and is likely to grow at a very rapid pace.
- There is a strong need for economical robots in mass production, as well as for increased speed to match that of dedicated, single-function equipment.
- Low and medium-volume assembly calls for robots with better performance to enable these to perform a large variety of different tasks.
- Peripheral equipment such as hands and automatic hand change units need to be more fully developed.

There are still many other areas of development vital to the robotisation of assembly lines. The authors hope to continue developing robot-based assembly systems and help the progress of factory automation.

References

[1]   Yonemoto, K. 1981. Industrial robots : Current trends and future prospects. *Robot*, 32 : 5-16 (In Japanese).

[2]   Hirabayashi, H. et al. 1983. Travelling assembly robot. In, *Proc. 13th Int. Symp. on Industrial Robots and Robots 7*, 17-21 April 1983, Chicago, Vol.2, pp.20.19-20.31. Society of Manufacturing Engineers, Dearborn, MI, USA.

[3]   Niinomi, T. et al. 1983. Heavy parts assembly by balancer-assisted robot system. In, *Proc. 15th CIRP Int. Seminar on Manufacturing Systems*, 20-22 June 1983, Amherst, USA. CIRP, Paris.

[4]   Ohashi, T. et al. 1983. The development of automatic assembly line for VTR mechanisms. In, *Proc. 15th CIRP Int. Seminar on Manufacturing Systems*, 20-22 June 1983, Amherst, USA. CIRP, Paris.

AUTOMATED ENGINE ASSEMBLY AT SAAB-SCANIA

S. Ericsson
Saab-Scania, Sweden

First presented at the 14th International Symposium on Industrial Robots and 7th
International Conference on Industrial Robot Technology, 2-4 October 1984, Gothenburg,
Sweden. Reproduced by permission of the author and IFS (Conferences) Ltd.

Scania's commitment to industrial robots and the new manufacturing
technology is a logical outcome of the company's desire to rationalise
manufacture, maintain progress in the field of product development at a
high level, and satisfy market demand for new products on the basis of
flexible manufacturing resources. The first industrial robots to play a part
in the assembly of Saab car engines were installed in the summer of 1983.
This coincided with the introduction of a completely new system for the
assembly of car engines, featuring automatic assembly carriers and the
application of industrial robots to assemble fly-wheels, end-plates,
timing covers and cylinder heads.

Saab-Scania is one of the largest industrial groups in Scandinavia employing
more than 40,000 people with an annual turnover in 1983 of about SEK 20.8
billion. The main part of the corporate operations lies in the field of
transportation and manufactures heavy trucks, buses, diesel engines,
passenger cars and aircraft. Manufacturing operations are conducted by
three corporate divisions with numerous subsidiary companies. The Scania
Division, the largest division in the Group, features heavy commercial
vehicles and diesel engines on its manufacturing programme, accounts for
roughly half the gross turnover of Saab-Scania and employs half of the total
Group workforce. The Scania Division is also responsible for the design,
development and manufacture of engines and transmissions for Saab cars.
  Scania's commitment to industrial robots and the new manufacturing
technology is a logical outcome of the company's desire to achieve optimum
rationalisation of the manufacturing process and to offer versatile products
designed to meet the specifications and demands of customers all over the
world for efficient haulage operations. Furthermore, its policy is one of
high-level product development aimed at getting the results of research work
out and onto the market using flexible manufacturing resources.

Given the diversified nature of the product manufacturing operations, industrial robot technology offers excellent opportunities for promoting rational and flexible operations virtually throughout the entire manufacturing process. The new technology has also enabled the elimination of poor work environments by promoting job satisfaction and optimising the job quality on the shop floor.

At the Scania Division, the first industrial robots were commissioned in 1976 to serve machine tools, and this is still the commonest application of industrial robot technology. Later in the same year the first process robot was commissioned to carry out welding operations; this was followed by the installation of 15 similar industrial robots. Today, process robots are used for spray-painting, deburring, the application of sealing compound, and other work. At present, about 85 industrial robots are in use.

In 1983 the use of industrial robots was extended following the commissioning of a series of units to assemble petrol engines for Saab cars. At present, nine industrial robots are used for this assembly work.

In conjunction with the transfer of petrol engine production to the Scania plant in Södertälje, the company adopted a completely new assembly system – group assembly. This system consisted of six assembly groups, normally with four operatives in each group, assembling virtually the entire engine, with the exception of a number of preassembled units. The operatives in each group were free to decide for themselves whether they wished to perform only certain steps in the assembly process or build the entire engine by themselves.

However, this assembly system did not prove entirely satisfactory – factors such as stress began to appear when operatives were working at different rates and there was a mutual feeling of harassment. Some of the work postures and lifts imposed too much strain on operatives, most of whom are women. Again, some of the tools used in assembly work, especially nutrunners, gave rise to physical discomfort in the form of backache and neck pains.

These factors in combination with several upward adjustments of the annual output rate – particularly at the beginning of the 1980s – and an accelerating process of wear and tear on existing assembly plant and equipment, triggered off a process that led to plans for the automation of trouble spots in the shop workfloor environment and for the creation of a more flexible production facility. Throughout 1982, a completely new assembly system was developed using industrial robots and sophisticated peripheral equipment to eliminate heavy lifts and monotonous work by creating a more pleasant work environment for our shop-floor operatives.

This automated assembly system was commissioned in the middle of 1983. Since then, further refinements to the equipment have been made and in the summer of 1984 several more industrial robots were commissioned to carry out assembly work. There is little difficulty involved in adding extra units as the company has configured all industrial robot groups to function as independent units and they are not linked to one another in any way. Configuring the assembly system in this manner makes it easy to withdraw a robot group and replace it with another model if and when such action is necessary. The company did in fact implement such a change last summer on one section of the assembly line when a further increase in the annual production rate necessitated the introduction of larger industrial robots to handle heavier, multispindle nutrunners. Industrial robots can be recycled and easily put to work on other sections of the assembly line.

In the design and development work involved on this new assembly system, special simulation software developed by Scania was used. By running the software on computer it is possible to simulate different production systems to identify availability levels, overall system capacity, buffer sizes, buffer locations, and many other factors.

When the Scania Division first began building Saab car engines in 1971, annual output was running at 50,000 engines in two models. Today the annual production rate is 110,000 engines in about 30 different versions. The existing production system is dimensioned for an annual output of 135,000 engines.

The assembly process

Automated section

The first section of the engine assembly line consists of a preassembly area using a conventional four-edge line. At this stage basic engine components such as crankshafts, con rods, pistons and big-end bearings, most of which are components common to different engine versions, are installed. Roof conveyors bring engine blocks up to the four-edge line. The engine serial number is stamped on the block at the last station in this section.

Fig. 1   Two IRb6 industrial robots thread and torque six different types of bolts used in the assembly of the end-plate and timing cover

The engine is then lifted onto a buffer conveyor where it is collected by an unmanned assembly carrier and transferred to one of the five manual stations that form what is known as basic engine assembly area I. Dowels, sleeves, oil filter with intermediate piece, gaskets, sealing compounds and other components are installed at these stations followed by the fitting of the end-plate and timing cover. The timing chain and its tensioner and gear are also fitted at these stations. The carrier then returns to collect the engine and take it to a transfer station where it is loaded onto a pallet and placed on the buffering roller conveyor.

This roller conveyor carries the engine to the second section of the basic engine assembly line commencing at an industrial robot station with two IRb6 units which install and torque the bolts securing the end-plate and timing cover (Fig. 1). The end-plate is fastened using a total of 13 bolts of two different lengths. These are fed to the robot by a centre-board feeder controlled by the robot control system. The timing cover is fastened with 10 bolts of four different lengths, also supplied by a centre-board feeder. Bolts are threaded and torqued using two-spindle nutrunners. For each of these operations, ASEA S2 control systems with seven axes of freedom of motion are used. The sixth and seventh axes of motion are used to change the centre-to-centre spindle distance and for turning the spindles in the horizontal plane.

The robots at this station use two separate software programs since the end-plate and timing cover are produced in two different versions, depending on engine. A fixed inductive transducer switches software by sensing the location of the water pump to identify the engine type.

Fig. 2   The 10 bolts securing the cylinder head are installed by an IRb6 that picks the bolts one by one from the hoist feed

Fig. 3 Using its three-fingered grip, the robot picks the flywheel straight off the pallet

Bolts are tightened to the specified torque references and the torquing process is checked to ensure that bolts are securely seated. This check is performed by measuring the extent to which a bolt protrudes. Should an error condition arise, the robot will output a signal to the sentinal tell-tale on the transfer pallet. This sentinal memory will be sensed immediately upstream of the adjustment station located further down in the system. The identification of an error condition on three consecutive engines will inhibit the third engine from leaving this station and will be followed by a shut-down and the triggering of a visual error condition annunciator. An additional check feature has also been incorporated in this station by setting a time limit for the assembly work. Exceeding the time limit will also trigger a visual error condition annunciator.

Engines that pass quality control proceed to the next station where a mechanical lifting device is used to fit the cylinder head and gasket manually. At the next station the ten cylinder head bolts are fitted by an IRb6 (Fig. 2), also using two separate software programs. Here, too, an inductive transducer is used to sense engine type and switch software between 8- and 16-valve engine versions. Although both engine types have the same bolt-hole configuration, different bolts are used.

In its first pass, the robot picks four bolts at a time from elevator feeders for each bolt length. The robot makes two additional cycles to pick up the rest of the bolts. The robot is programmed to pick several bolts at a time in order to minimise the number of return travel cycles and keep station dwell times as brief as possible. The two sets of control software differ in that there is a 'loading position' for each elevator feeder and the movement pattern for the installation of bolts in the 16-valve cylinder head is somewhat different.

This is partly due to the fact that the twin, overhead cam shafts entail a somewhat more complicated 'unloading position'.

The cylinder head bolt torquing operation is performed at two separate stations – bolts are pretorqued to about 90Nm at the first station using a ten-spindle nutrunner with an integrated torque control device. The flywheel installation is inserted at the next station to allow the cylinder-head gasket to bed down after the pretorquing operation.

An IRb60 with a three-clawed gripper picks the flywheel directly from pallets located on a turntable (Fig. 3). Flywheels arrive on pallets and are located in a specific picking pattern on the pallet spacer inserts. The robot features an inductive transducer to sense pallet position and identify one of the 11 levels from which it should pick the flywheel. The robot then grasps the flywheel and lifts it up onto an orientation station where the ignition timing pin is used to position the flywheel correctly to mate with the dowel on the end of the crankshaft.

After installing the flywheel, the robot withdraws from the workpiece to pick a new flywheel while a four-spindle nutrunner mounted on tracks running beside the roller conveyor moves in to torque four bolts before indexing to 45° to tighten the remainder. The nutrunner incorporates an automatic torque and angle of rotation check feature. The robot is programmed to run on four separate sets of software because different engine types are equipped with a total of four different versions of flywheel and flex plates – the latter for automatic transmissions. Once a station has been loaded with a specific variant of flywheel or flex plate, the operative keys in the number of engines to be equipped with this component and an automatic count-down to zero follows, whereupon the unit switches to a fresh program. The nutrunner is reprogrammed simultaneously because two different lengths of bolt are used.

At the next station, the bolts are retorqued to the yield limit by a microprocessor-controlled, fully automatic, ten-spindle nutrunner incorporating an automatic torque and angle of rotation check feature. A buffering conveyor then transfers Saab 90 and 900 engines to a transfer station where they are lifted onto loop-controlled assembly carriers for further transfer to the 23 manual final assembly stations.

In contrast to other engine variants, the new Saab 9000 engine is equipped with an oil pan. All 9000 type engines proceed directly to the next assembly station where they are lifted up by a gantry hoist and secured in position before moving on to a manual station where the operative inverts the engine to bring the bottom to a convenient position in which to work. An oil suction pipe is fitted to the block at this station at the same time as the oil pan is inserted under the engine, and its contact surfaces are coated with gasket compound. This operation is performed at a robot station to the side of the assembly line using a SCARA robot.

After the installation of the oil suction pipe, the hoist turns the engine back to the vertical position and lowers it above the oil pan on the pallet, before it moves to a 17-spindle nutrunner that secures the oil pan from below. The engine then moves to a transfer station and is lifted onto a loop-controlled assembly carrier. The empty pallet receives a new oil pan and moves forward to the SCARA robot for application of gasket compound.

Just described is an improved version of the assembly system commissioned in 1983. On the basis of the experience gained from this assembly system, stage two in the current programme for the rationalisation of petrol engine assembly has now been implemented. The company has been able to create a flexible production facility that can adapt quickly to accommodate new engine types and rapidly step up production rates because it opted right from the beginning for a decentralised control configuration for individual sections of the assembly system with each robot controlling its own peripheral equipment.

Final assembly

Engines are then picked up by loop-controlled assembly carriers for transfer to the 23 manual assembly stations that constitute fitting-out assembly areas I and II. On arrival at an assembly station, the engine brings with it a picking bin containing most of the components that will determine the ultimate version of the engine. The first section of the fitting-out assembly area contains eight workstations where fuel pump, belt pulleys, thermostat housing, spark plugs and valve cover are installed. By using vertically adjustable and pivoting assembly carriers, tiring work postures have been successfully eliminated. Station dwell times range from 7 to 11min. This section also features three workstations exclusively devoted to the assembly of the turbo unit. The loop-controlled assembly carriers transfer engines to the second section of the fitting-out assembly area which contains 12 workstations. Components fitted here include distributor, ignition cables and a variety of pipes and hoses for water, fuel, vacuum and emission control systems. Station dwell times in this section vary from 7 to 18min, depending on the amount of work involved in the particular engine version.

On leaving this section, the engine is ready for test running. Still on the loop-controlled assembly carrier it is transferred to one of the 16 test cells in the engine test area. The engine test cycle including setting up and tearing down lasts about 20min. After test running, engines pass through a cooling tunnel before arriving at four manual stations where they are fitted out for shipment. This process includes installing a number of minor parts and a variety of hoses and plugs that cannot be installed until the engine has been test run. The engine is then packed in a plastic bag before transfer to a buffering station to await consignment to the car assembly centres in Trollhättan, Arlöv and Uusikaupunki (Finland).

The final assembly system uses 90 loop-controlled assembly carriers under microprocessor control. This loop system can be easily expanded because each microprocessor only controls a limited section of the system. The microprocessor systems used in the automated section of the assembly process and in the final assembly system are not linked to any mainframe computer. Nevertheless, the system has the capability for such a link up should the company, at a later date, wish to collate data for assessing production performance, equipment MTBF and similar factors.

Cylinder head manufacture

With the launch of the third generation of Saab turbo engines in 1983 – the 16-valve power plant – the company realised the need to increase its cylinder head manufacturing capacity. The new cylinder head marks a radical departure from the 8-valve engine; firstly, in view of the fact that it has 16 valves and, secondly, through the adoption of twin, overhead cam shafts. The company opted for the solution of relocating the machining and assembly of the new cylinder head, as well as the assembly of cylinder heads for other versions of the Saab engine, to a completely new workshop.

For processing the new 16-valve cylinder head, a completely new machine tool park, including three industrial robots for conventional machine tool feed as well as deburring work, was purchased. The 16-valve cylinder head also differs in the context of assembly work and separate assembly lines have been set up for both types of cylinder head. Valve-spring assemblies with associated valve collets were assembled by hand in the past. This is very monotonous work and the company believed that it should be performed by robots, particularly in view of the fact that the 16-valve cylinder head uses twice as many parts as a conventional unit. This would require even more manpower to keep assembly times at a minimum and would still not have eliminated the monotony. Using industrial robots with very sophisticated peripheral equipment, automation of all of this work has been achieved.

After machining, cylinder heads are transferred to the assembly line and lifted onto a buffering, motorised transfer conveyor. The cylinder head is first blown clean and then transferred to a turntable unit where it is inverted to permit the installation of the valves. This work is performed manually.

Fig. 4   Robot installs valve-spring assemblies on the automated section of the cylinder head assembly line

The cylinder head is then inverted once more and transferred to the next station where an IRb6 installs the valve-spring assembly consisting of bottom washer, valve spring and top washer (Fig. 4).

A vibration feeder and centre-board feeder carry washers and springs to a valve-spring assembly station where the units are stacked to form packages of 8 and 16 units, depending on engine type. Spring assemblies are stored in a magazine against which the robot 'docks' using its gripping device/magazine. The robot then proceeds to assemble the valve-spring units one at a time.

The IRb60 at the next assembly station may, at first glance, appear substantially overdimensioned for the job of assembling valve collets that scarcely weigh 10g. Nevertheless, the size of this industrial robot is more than justified. Two collets are used to secure each valve; these are fed the right way round on a vibration feeder to a magazine. The robot moves forward and picks up the collets to transfer them to its magazine and also features a tool for compressing valve springs. This is the ·operation that justifies the size of the robot, as a force of about 750N is required to compress the valve spring, enabling the collets to fall into place.

The sophisticated tool on the industrial robot incorporates a measuring device to ensure that the collet is correctly installed by checking the height of the top washer. If a collet has been incorrectly installed, there will be a discrepancy in the washer height. This is input to the computer controlling this section of the assembly line. Before such a cylinder head reaches the next station, it will be automatically diverted to an adjustment station.

The two assembly lines then merge into a single section with two automatic stations. One is the station where the valves are operated to seal against the valve seat and the other is a station where a leak test is conducted. The assembly process then continues along two separate lines where other components in the valve mechanism are installed together with the inlet and exhaust manifolds.

Another robot is involved in the cylinder head assembly process. This is a SCARA robot that installs a valve stem seal on the upper section of each valve guide after the 16-valve head has left the process line. Seals are fed up to the robot tool by a centre-board feeder.

Concluding remarks

It is evident that robots have been used to perform very advanced operations in the new engine assembly system. In fact, some of these operations have never previously been performed by robots. Nevertheless, Saab-Scania still has a long way to go in developing assembly techniques based on the use of industrial robots. In general, although the company is very pleased with the performance of the equipment installed, it is closely monitoring developments in the industrial robot sector and has great expectations of future generations of industrial robots.

3

Product Design for Robot Assembly

As our experience of robotic assembly accumulates, the evidence in support of a serious approach to product design becomes unavoidable. Product design is, frequently, the key to success and profitability in robotic assembly applications.

PRODUCT DESIGN FOR SMALL PARTS ASSEMBLY

J. Browne and P. O'Gorman
University College Galway, Ireland
and
I. Furgac, W. Felsing and A. Deutschlaender
Fraunhofer - Institüt für Produktionsanlagen und Konstruktionstechnik
(IPK), West Germany

Based on the design concepts outlined in the introduction of this paper, a
series of interrelated design rules for robot based assembly can be
identified and will prove industrially useful. Using electronics assembly
as an example industry, these design rules are proposed and outlined.

Application of robots in industry has largely been confined to relatively
primitive tasks – machine loading/unloading, spot welding, spray painting,
etc., whereas relatively few applications in assembly have been realised.
Researchers have adopted two main approaches: the development of
assembly robots and the redesign of products for robot assembly. The first
approach involves the development of 'intelligent' sensor-based robots with
sufficient accuracy, speed and repeatability, 'universal' grippers and capable
of being programmed in task-orientated robot languages, usually based on
principles of artificial intelligence. A major stumbling block here is the need
for massive computational/processing power far beyond todays generation
of processor-based robot controllers. Even the availability of distributed
systems based on 16-bit microprocessors is highly unlikely to change the
situation dramatically.

Both approaches are perfectly valid and somewhat complementary.
However, it is the authors' view that in the medium term the second
approach is likely to be fruitful from an applications perspective. A
difficulty, however, with the design for assembly approach is that its major
emphasis seems to be on the redesign/rationalisation of existing products/
assemblies and subsequent application of the 'lessons' learned to the design
of new products.

If we consider an industry where flexible automation has been achieved,
an alternative approach to the problem can be recognised – in fact an
alternative methodology. Stuffed printed circuit boards (pcbs) for use in
such products as computers, telecommunication equipment, control

equipment for machines (e.g. CNC controls) are assembled in batches. Yet the vast majority of the components to be assembled, resistors, body formed capacitors, integrated circuits, are assembled using computer automated assembly machines. Such machines include variable centre distance machines (VCDs), fixed centre distance machines (FCDs), which are in both cases 'linked' to sequencing machines, and also dual in-line package (DIP) insertion machines.

In the authors' view, two key concepts have allowed computer automated assembly to be achieved in the electronics industry. One is the separation of 'function' and 'form' in the design phase and the second is the use of standard components – that is, standard from an assembly perspective.

The circuit design (function) and the pcb layout (form) are effectively separated. The 'form' stage of the design involves the layout of standard components on standard sized boards. Resistors, for example, are supplied in reels with fixed distances between individual resistors on the reel, and fixed 'end-to-end' lengths irrespective of 'resistance' nominal values.

Applications of robots in assembly

The assembly of discrete parts is an area of batch manufacturing where productivity has lagged behind that of other processes such as machining and inspection. Growth in manufacturing productivity has largely been due to the substitution of power driven machines and new technological methods for manual labour. This is particularly true of machining and forming processes and of materials transfer systems. Conceptually, programmable automatic assembly is analogous to numerically controlled (NC) machining. However a major difference between the economics of NC machining and programmable automatic assembly is that in conventional machining processes the machining is performed by manual operation of a machine tool, whereas manual assembly is performed by workers using simple tools. Thus, automation of the machining process requires *only* to automate existing machinery, whereas in programmable assembly an entirely new machine must be designed and built which can compete effectively with human operators.

To date there have been few examples of the application of robots in light mechanical assembly in industry. The major difficulty, in simple terms, is that it has not been possible to develop a robot which can compete effectively with the human hand. However, there has been a small number of successful implementations where users have recognised the need for product design for assembly[1-7] (See also Chapters 1 and 2).

Robot design for assembly

This section outlines the approach adopted by several robot manufacturers in the design of industrial robots for assembly.

DEA PRAGMA A3000

The robot arm is characterised by a Cartesian mechanical structure (three linear axes), expandable by the addition of rotary axes up to a total of five degrees of freedom (see page 3). The Cartesian structure is the most suitable for resolving assembly problems; it allows precise movements in directions in which fitting and joining operations are made most frequently.

The 'control unit' is able to control up to four arms, monitor the sensors and grippers, and to interface with all the external equipment normally fitted in an automatic assembly system.

YESMAN

The YESMAN robot has been designed around the premise that a significant increase in productivity and quality can be obtained by a combination of human and machine skills (see page 39). This is achieved by designing YESMAN such that it can work safely alongside human operators.

The advantages of a 'mixed' assembly process are several. The first has already been mentioned, namely that cost is not built into the machine in trying to make it do a task for which it is inherently unsuited. Secondly, the implicit quality control that is readily achievable with manual assembly is retained. Thirdly, the combined attitudes of manual and machine assembly (flexibility, precision, quality and speed) can be greater than with either alone.

Among the factors influencing the sale and usage of assembly robots is the need for design changes. In order to permit pure robot assembly to take place it is often necessary to redesign the product to make the assembly task easier. The general quality and precision of piece-parts often has to be raised.

The YESMAN robot has been designed to carry out a large range of assembly tasks, including basic pick and place (but not bearing fit insertion), self-tapping screw insertion, rivetting, adhesive and sealant application, soldering, snap connection, circlip and other fastener applications, adjustment of rotary or linear controls, and heat or ultrasonic bonding. The YESMAN is essentially a workstation and comprises:

- Twin robot arms which can work either synchronously or independently.
- An integrated workbench which incorporates jig mounting features, tool holders, basic operator interface and data carriage reader.
- An integral control computer which can be used either to control the robot or to program it.

The workstation has a working area of 550×250mm and assumes that the workpiece will have a reasonable level of vertical access. The working volume extends this area vertically by 250mm.

The working volume is served by two servo arms (although the machine can be configured as a single arm unit if required). Each arm has three servo axes powered by low-cost dc motors. The maximum arm load is 1kg although heavier loads can be accommodated with some loss of accuracy. The maximum speed is 30m/min and positional accuracy is ± 0.3mm with a repeatability of ± 0.1mm.

The control system for YESMAN involves a supervisory computer overseeing three microcomputers, one for each of the three servo axes. In addition to overall supervision of the robot, the supervisory computer needs to take data relating to a task and convert it from Cartesian (x, y, z) coordinates into joint coordinates ($\oplus$, θ, z) in real time. Furthermore, a number of interface options and I/O ports have been designed into the control system to allow YESMAN to interact with other elements in the workstation. The robot may be programmed either through 'joystick learning' control or via a high-level interactive language.

IBM RS-1

The design of IBM's RS-1 (now termed 7565) gantry type robot has been described elsewhere[8] (see also page 81). It consists of a rectangular frame supporting a hydraulically driven arm assembly, which, in turn, consists of three linear actuators and a gripper with linearly actuated fingers. The three linear actuators are arranged orthogonally to carry the wrist while providing independent motion in x, y and z directions. The three rotary actuators of the wrist control is the angular orientation of the gripper in roll, pitch and yaw.

For assembly applications, the orthogonal structure has several advantages over anthropomorphic style manipulators. Assembly applications frequently involve many short motions, the durations of which are limited primarily by robot accelerations. A Cartesian box frame robot can be made quite rigid, allowing it to be operated with high accelerations.

The rigidity also improves the robot's precision, which is important in assembly. Furthermore, the orthogonal structure makes the precision essentially independent of the robot's position. In addition, it decouples the dynamics of the x, y and z axes, allowing them to be controlled largely independently. The orthogonal structure also allows considerable modularity in constructing robots of different sizes and shapes. The same linear joint is used on each axis, and the axes can be built to different lengths.

In many assembly applications, only three translational degrees of freedom are required. With an orthogonal structure, three linear joints can provide this freedom, whereas a robot with only revolute joints needs at least five joints to provide the same freedom. Furthermore, in applications requiring only three degrees of freedom, Cartesian coordinates are identical to 'joint angle' coordinates, eliminating the need for subroutines that perform coordinate transformation computations. When multiple arms must work in coordination at the same workstation, collision avoidance is an important issue. With an orthogonal structure, multiple arms can share a linear axis, reducing the cost and simplifying the collision avoidance computation to a single degree of freedom.

SCARA

The SCARA (Selective Compliance Assembly Robot Arm) is an assembly robot developed in the late 1970s and early 1980s by Professor Makino of Yamanashi University, Japan (see page 13). Results from experiments in assembly research indicated that for some directions stiffness was required while for other directions compliance was required. The SCARA robot was designed to meet these requirements.

The SCARA has four degrees of freedom:

- Horizontal rotation of the joints θ_1 and θ_2 deciding planar position of the tool point.
- Another rotation θ_3 deciding the orientation of the tool about the z-axis.
- The last degree of freedom θ_4 for the vertical motion of the tool.

This kind of construction is suitable for the assembly task of 'down-to-up' mounting which is considered to form 80% of the existing work if the product design is suitable. Several features of the SCARA which facilitate assembly are:

- Selective compliance – the robot can adapt to small variations in position.
- Motion control – smooth but rapid moves are possible.
- Payload – the SCARA can hold a large payload because of the stiff structure in the z-direction.
- Working areas – the SCARA can cover larger working areas than a Cartesian coordinate robot of the same size.

The design for assembly approach

The assembly process can be subdivided into the following operations: feeding, handling, mating, inspecting, special operations (e.g. deburring), and adjusting. The basic problem is how to design products to facilitate these operations.

One approach to the problem is to reduce the number of parts in an assembly. In determining the theoretical minimum number of parts needed in an assembly, Salford University have developed four criteria to decide whether or not a part may be combined with another. If a part meets any one of the criteria it must remain as a separate part. The four criteria are:

- The part (or subassembly) moves relative to any other part in an assembly during the normal function of the assembly.
- The part must be made of a different material from its mating part (for insulation, vibration, damping, etc.).
- The part needs to be disassembled for servicing.
- The part, combined with its mating component, would prevent assembly of other parts that meet the first three criteria.

According to Treer[9], there are three fundamental approaches to designing a product for automated manufacturing:

- Design for simplification.
- Design for ease of automation.
- Design for modular construction.

Design for simplification is based on simplifying the overall manufacturing process. Often, new product designs include off-the-shelf items, standard items, or parts that are possible to make with a minimum of experimental tooling. This approach can constitute a major simplification of the total manufacturing process. There are a number of factors to be considered regarding the cost-effectiveness of different product designs:

- Use of standard components or materials.
- Use of sophisticated manufacturing processes not available to the prototype builder.
- Minimisation of inventories of components and materials.
- Reduction in handling requirements.
- Reduction in malfunctions and down time.
- Reduction in opportunities for errors.

Designing for ease of automation is concerned with the general concepts and design ideas which will help simplify the automatic parts feeding, orientating and assembly processes. It is important to design products to be machine assembled from the top down and to avoid forcing machines to assemble from the side or particularly from the bottom. The ideal assembly procedure can be performed on one face of the part with straight, vertical or horizontal motions, keeping the number of faces to be worked on to a minimum.

Product design for modular construction has two purposes. The first is improved efficiency which may be achieved by standardising subassemblies. This standardisation results in increased production volumes of fewer different components and in reduced inventories. The second reason is that many assemblies are far too complex to make complete in one pass, therefore it is more efficient to break down the total assembly into a series of subassemblies.

A more systematic approach to designing products for robotic assembly is given in the next section of this paper. However, it is useful to quickly review an industry where many of the above ideas have been applied in practice. The electronics industry is a prime example of the benefits which can accrue from the design for assembly approach.

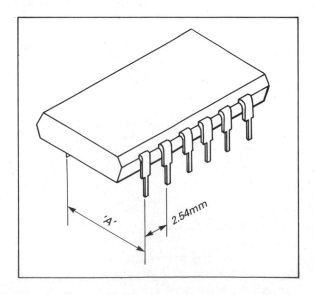

Fig. 1   Dual-in-line package (DIP) configuration

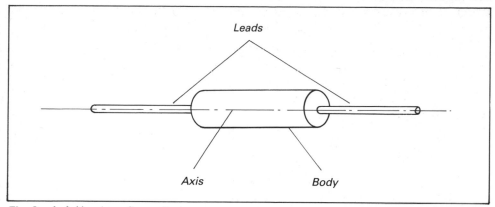

Leads

Axis

Body

Fig. 2   Axial lead configuration

A large element in manufacturing in the electronics industry is assembly of printed circuit boards (pcbs). The design for assembly approach has yielded many benefits, including reduced manufacturing costs and increased quality. One major approach to design in the electronics industry has been to decrease component size while increasing the power and versatility of the components. This is, in effect, the 'designing-out' of assembly. An example of this is in the use of dual in-line packages (DIPs), where many functions may be included in the one chip thus reducing assembly requirements to the insertion of one component. With this comes the other important advantage of greatly reducing the physical size of components.

Standardisation has been applied to DIPs in two forms. The first form is in electronic functioning, meaning that the 'same' integrated circuit (IC) is available from several suppliers. The second form of standardisation is DIP format. The format is specified by several engineering standards, namely DIN 41866 (1973), 191 IEC 1-50a (1976), and JEDEC (1977). These are in effect very similar and the JEDEC standard has become very popular, having been adopted by many DIP manufacturers.

The DIP format is illustrated in Fig.1. The leads have a 0.1 inch pitch spacing along the side of the body. The 'between centres' distance (A in Fig. 1) is generally one of either 0.3, 0.4 or 0.6 inches. Of course, the number of leads depends on the complexity of the IC, but typically lies between 2 and 40. Discrete components (e.g. resistors) have also been standardised to a large degree, having an axial format (Fig. 2).

The standardised format simplifies the insertion process as parts variety has been decreased. Furthermore, parts feeding is facilitated by component design. DIPs are supplied in standard slide magazines, and axial components are supplied in bandoliered form as illustrated in Fig. 3.

At present, about 80% of the components on a typical pcb can be assembled automatically. The remaining 20%, the non-standard components, have still to be hand assembled. Many large companies, e.g. Westinghouse, are now examining means of automating the assembly of these components. The redesign of these components has been identified as a major requirement for automation[10]. This is an important area of

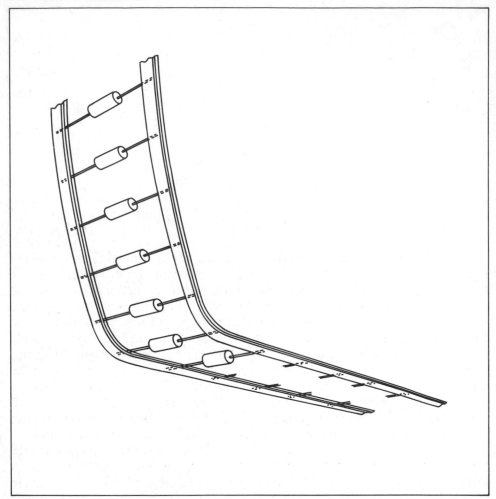

Fig. 3   Bandoliered components

research from a manufacturing point of view as the cost of assembly of the non-standard components is generally greater than that for assembling the standard components. This is despite the fact that the non-standard components account for only one-fifth of the total components on the board.

The assembly task itself, although demanding high positional accuracy, is facilitated by the fact that the direction of assembly is unidirectional. Components are generally assembled in a direction perpendicular to the plane of the pcb. It is expected that robots will play an important part in future pcb assembly systems.

A systematic design procedure for robot orientated assembly

As mentioned previously it is essential that products be designed for robot orientated assembly. A systematic procedure for this is illustrated in Fig. 4. This is an iterative design procedure and not a straight-through process. Each of the three stages is now considered in detail.

Fig. 4   Functional procedure

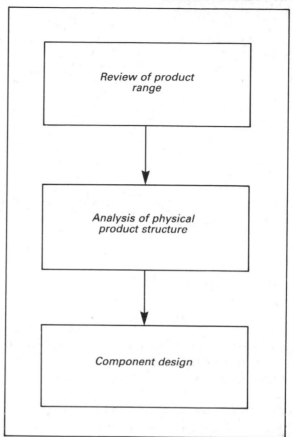

Review of product range

A large product range and a large variety of product styles require a high degree of flexibility in the assembly system. Generally it is true that the greater the flexibility desired the more expensive the assembly system becomes. This could result in a system that is outside the bounds of economic and technological feasibility. Among the objectives of this review of product range are to ensure high utilisation, high productivity and consistent high quality [11,18].

Utilisation is influenced by the size of the production run assembled on the same equipment. Hence, the designer should avoid variations or ensure that variants can be assembled in the same way. The productivity of the system is a function of the number and length of stoppages. Again there is a trade-off between an extremely flexible plant that can accept many variations in components and a range of components with few variations.

Possible ways of satisfying these objectives are to reduce/rationalise the product range or to examine the commonality of parts and subassemblies over the product range, as shown in Fig. 5, and try to increase the degree of commonality. It is not just production cost savings that accrue from a rationalisation of product range, there are also organisation savings resulting

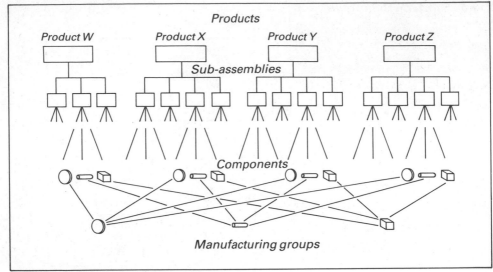

Fig. 5   Parts commonality

from possibly smaller product range and/or fewer parts in inventory, fewer parts drawings. Further savings will result from reduced programming, reduction in the number of robot stations, and reduction in the range of robot grippers.

Analysis of physical product structure

The product structure is the outcome of the quantitative structure level design. This stage of the functional procedure is therefore concerned with the design of product so that robot assembly is facilitated. The product designer works on two levels, a fundamental structure level where techniques and solutions are logically connected, and the quantitative structure level where decisions on distances, tolerances, positioning in space and division into machines parts are made[12].

To achieve a product design that is suited to robot orientated assembly, Andreasen et al.[18] recommend that various product structures be systematically examined as an aid to alternative design. The alternative product structures are:

- The frame – one basic component carrying all other components.
- Stacked assembly – the components are assembled by stacking them on top of each other and secured by, usually two, surrounding components.
- Composite product – different materials are combined to meet different demands.
- Base component-product – the base is used for assembly and transport.
- Product created from modules – a product composed of larger self-contained parts or functional units with simple relations with the rest of the product.
- Building-block system – a system of products which are structured in such a way that all products can be constructed of a number of building blocks.

There are two general but useful design principles at this stage: design for simplicity and design for clarity. Two techniques that have proved valuable and that should be adopted are:

- Separate function and form.
- Integrate and differentiate.

The first technique, separation of form from function, has been very successfully applied in the electronics industry where the function, as exemplified by a circuit diagram, has been developed separately from the form, i.e. the pcb layout and physical component selection.

At the form stage of the design there are specific rules regarding product design. For example, the axis of the components must be orientated in a north south or east west direction. This design is facilitated by the large degree of standardisation in electronic components.

An example of the application of the integration and differentiation principle is given in Fig. 6. Integration results in fewer parts than differentiation.

One method of reducing the difficulty of the assembly task is to reduce the content of the task by using fewer parts. However, fewer parts may result in more complex moulded or machine parts with a consequent increase in their cost of manufacture. Therefore there is a need for a model to help evaluate these costs so that the total can be minimised. According to Boothroyd[14], reducing the number of parts in an assembly means combining several components into one; yet parts costs still decrease in spite of increased parts complexity. Some of the rules that should be followed at this stage of the functional procedure are outlined below[11,13,15,16,18]:

- Avoid separate connecting elements.
- Standardise joining points. If different product styles are needed then all the differently shaped components should have the same connecting points and assembly method. This reduces the need for tooling and set-up.

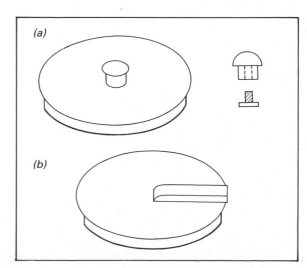

Fig. 6 Design integration and differentiation: The lid (a) consists of three parts which have to be assembled. However these have been integrated into a single plastic moulded part (b) which does not require any assembly

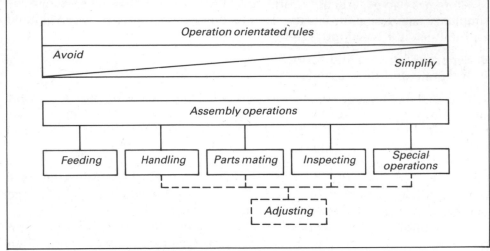

Fig. 7   Direction of assembly-orientated design

- Avoid separate connecting elements.
- Replace screwed connections by locking connections. Screwing involves numerous individual steps – ordering of screw, setting-up, singling out, feeding, locating, etc.
- Reduce the number of assembly parts – individual parts and design groups should be combined.
- Reduce the extent of the final assembly operation by defining new sub-design groups.
- Standardise assembly direction and joining by motion along a single axis.
- Avoid tight tolerances where possible.

Design of components

At this stage of the design procedure the central theme is the avoidance and simplification of assembly operations, as shown in Fig. 7. There are rules which should be applied to each of the assembly operations outlined in Fig. 7 in order to achieve design for robot orientated manufacture:

- *Feeding*
  Avoid orientation operations
    - use magazines
    - use bandoliered components
    - integrate the production of components into the assembly, e.g. springs
    - avoid tangling/nesting
  Facilitate orientation operations
    - avoid clamping or hooking
    - put special faces in the component for orientation
    - avoid components of low quality
    - make the component symmetric
    - or make it clearly asymmetric

As described previously, in the electronics industry axial components are supplied in bandoliered form and integrated circuits are supplied orientated in slide magazines.

- *Handling*
  Eversheim[15] subdivides the handling operation into three stages: picking up, moving, and laying down. He advocates the simplification of the handling operation by: (a) the passive reduction of demands that are made on the handling device, by removing difficulty in the execution of handling functions; and (b) the improvement of the possibilities of handling a part by adding special form elements to the part, that support the execution of the handling operation actively. The requirement for handling is further decreased by the integration of components.

- *Parts mating*
  - put special faces on the component for guiding purposes, e.g. lead-in chamfers
  - make all joins simple
  - reduce the number of stop faces
  - insertion should be along a single axis
  - reduce the number of parts and connecting elements by integrating production methods.

For example, in automatic pcb assembly, the probability of successful insertion of IC leads is greater if the leads are pointed than if the leads are straight ended[17]. Printed circuit boards are a good example of products where insertions are along a single axis (vertical).

- *Inspection*
  - facilitate ease of access to test points

- *Adjusting*
  - design products to facilitate adjust operations

The design of the individual components may force the designer to return to the second stage of the functional procedure to reconsider the product structure.

Summary design rules

A summary of the design rules previously discussed is now presented. Essentially there are two design rules: avoid and simplify. This design methodology is shown in Fig. 8.

The rules presented below result from an application of these two basic principles to the area of light engineering assembly. Clearly manufacturers in other industries will have to develop their own set of rules.

The design rules fall into four categories or levels:

- Level 0 – Overall strategy
- Level 1 – Product range
- Level 2 – Product structure
- Level 3 – Component design

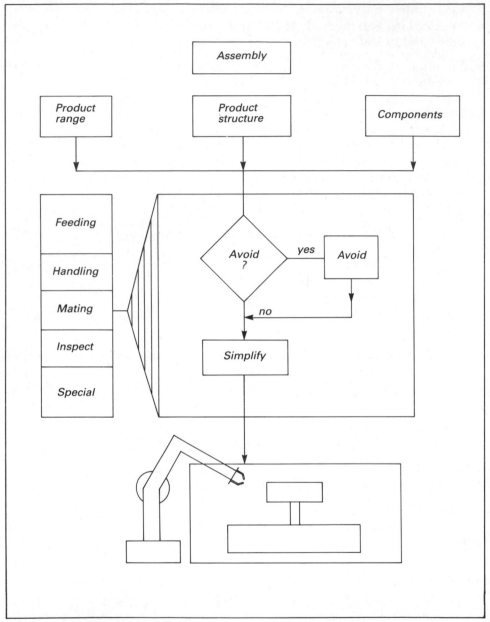

Fig. 8   Design rule methodology

Level 0 – Overall strategy
 1. Products must be designed for robot orientated production as well as for performance, reliability, maintainability, safety, aesthetics, etc. This requires the involvement of the production department at an early stage of the design process.
 2. Formulate a manufacturing/product strategy. Strategic decisions must be made about the manufacturing system and the product range.

Level 1 – Product range
1. Minimise, and where possible, avoid product variations, or ensure that variations can be assembled in the same way.
2. Rationalise the parts range within the products. This involves parts families, not individual parts.

Level 2 – Product structure
1. Use standard assembly modules where possible.
2. Design the product for simplicity and clarity.
3. Try to reduce the number of parts in the product. This reduces assembly costs but may increase parts costs. Use integration/differentiation principle.
4. Examine alternative product structures and use a single one through the product range if possible. Basic product structures include the frame, stacked assembly, composite product, base component-product, modular product, building block.
5. Avoid joins where possible. Reduce the number of joining operations in the assembly of a product.
6. Standardise joining points.
7. Avoid using separate connecting elements where possible. This also reduces the number of parts to be assembled. It means designing parts so that they can be joined together without the need for a separate part to hold them together.
8. Replace screwed connections by locking connections where possible. This eliminates parts (screws) and simplifies assembly.
9. Standardise the direction of assembly.
10. Avoid tight tolerances where possible.
11. Ensure that test, inspection and adjust points are freely accessible.

Level 3 – Component design
1. Design components for simplicity and clarity.
2. Use standard components where possible.
3. Avoid, if possible, or simplify parts feeding. This reduces the requirement for an expensive and complex array of parts feeding equipment.
4. Avoid orientation operations where possible.
5. Use magazined parts where possible.
6. Use bandoliered components where possible.
7. Integrate the production of components into assembly.
8. Avoid tangling, nesting of components.
9. Facilitate orientation operations.
10. Avoid clamping or hooking if possible.
11. Put easily identifiable orientating faces on components.
12. Avoid using low quality components.
13. Either make the components symmetric or else totally asymmetric.
14. Avoid, if possible, or simplify the handling operation. (Handling is a non-productive operation and should therefore be avoided.)

15. Simplify and reduce the handling required by the part. Special problems are presented by delicate parts or parts with a precision ground finish. Reduced handling requirements may mean fewer and/or less versatile grippers are required.
16. Design the part so that handling is facilitated. This may mean the addition of special surfaces that can be grasped by the robot gripper.
17. Simplify the parts mating process. Simplification of this operation can result in shorter assembly times and in a greater probability of successful assembly.
18. Put special guiding faces on the component.
19. Reduce the number of stop faces where possible.
20. Design components to facilitate inspection and permit free access to test and adjust points.

Concluding remarks

The question remains as to how the procedure outlined in the previous section could be realised in practice. In the short term, it is suggested that designers be 'educated' in these rules and that individual companies/industrial sectors refine the rules in context of their own experience/products. Several organisational changes are required to facilitate this. A major one is the involvement of the assembly department in the design process. This not only brings assembly experience into the design process but also acts as a channel for feedback on the ease of assembly of recently introduced products. Hence, the system, can learn from experience.

As product design is an ongoing process in a manufacturing organisation, it is likely that, in the future, each organisation would have access to a computerised design facility. The design rules will reside in this facility. From the software point of view it is likely that the rules will be implemented on the principles of artificial intelligence. A software generated line of questions and answers about features of the design would lead to the identification of undesirable features. This AI system would form the core of the procedure. Again, 'learning from experience' will constitute a very important function in the system. This computer-aided design (CAD) system will be tightly integrated with other computer systems in a computer integrated manufacturing (CIM) environment. For example, by interfacing CAD with computer aided process planning (CAPP) the sequence of assembly tasks could be simulated in 3-D graphics. Together with the AI base of design rules this would allow the examination of the assembly sequence and also highlight any problems resulting from the present design of the parts/products.

Among others there is a requirement for a manufacturing cost model. This would use current manufacturing cost data to assist the designer in the selection of processes and also assist in the decision making regarding the trade-off between 'parts complexity' and 'degree of assembly' required.

Design is a creative activity and the design procedure for robot orientated assembly should facilitate that activity. The system, by being user-friendly and performing many of the tedious calculations and alterations, will facilitate the design activity and facilitate the development of better products more quickly.

References

[1] Del Gaudio, I. et al. 1980. Matching the assembly robot to the factory. *Assembly Automation*, 1(1): 26-29.

[2] Hartley, J. 1980. How the Japanese put fuel gauges together in one second. *Assembly Automation*, 1(1): 47-51.

[3] Rodgers, R.C. 1982. Planning for automatic assembly. *Production Engineering* 29(2): 40-45.

[4] Lane, J.D. 1980. Evaluation of a remote centre compliance device. *Assembly Automation*, 1(1): 36-40.

[5] Automatic disc assembly. *Systems International*, 11(4): 41, 1983.

[6) Robot helps in disc drive assembly. *Electronics*, June 1982: 48-49.

[7] Badger, M.A. et al. 1983. The assembly robot: Applications in flexible assembly systems. In, *Proc. 2nd European Conf. on Automated Manufacturing*, pp.243-249. IFS (Publications) Ltd, Bedford, UK.

[8] Taylor, R. and Grossman, D. 1983. An integrated robot system architecture. *Proc. IEEE*, 71(7): 842-856.

[9] Treer, K.R.1979. *Automated Assembly*. Society of Manufacturing Engineers, Dearborn, MI, USA.

[10] O'Gorman, P. 1983. *Automatic assembly of Non-standard Components to Printed Circuit Boards*. M.Sc. Thesis, Cranfield Institute of Technology, UK.

[11] Andreasen, M.M., Kahler, S. and Lund, T. 1982. Design for assembly – An integrated approach. Principles and strategy. In, *Proc. 3rd Int. Conf. on Assembly Automation*, pp. 215-228. IFS (Publications) Ltd, Bedford, UK.

[12] Tjalve, E. 1979. *A Short Course in Industrial Design*. Butterworth, London.

[13] Boothroyd, G. and Dewhurst, P. *Design for Assembly – A Designer's Handbook*. University of Massachusetts, Amherst, USA.

[14] Boothroyd, G. 1983. Effects of assembly automation on product design. *CIRP Annals, Manufacturing Technology*, Vol. 32/2, p.511.

[15] Eversheim, W. and Muller, W. 1982. Assembly orientated design. In, *Proc. 3rd Int. Conf. on Assembly Automation*, pp.177-190. IFS (Publications) Ltd, Bedford, UK.

[16] Warnecke, H.J. and Walther, J. 1982. Automatic assembly – State-of-the-art. In, *Proc. 3rd Int. Conf. on Assembly Automation*, pp.1-14. IFS (Publications) Ltd, Bedford, UK.

[17] Hartmann, G. and Winckler, R. 1980. Insertion reliability at automatic assembly of pcbs and their influencing parameters. *Siemens Components*, XV (1980), No.1.

[18] Andreasen, M.M., Kahler, S. and Lund, T. 1983. *Design for Assembly*. IFS (Publications) Ltd, Bedford, UK.

PRODUCT DESIGN FOR ROBOTIC AND AUTOMATIC ASSEMBLY

J. F. Laszcz
IBM Corporation, USA

First presented at the Robots 8 Conference, 4-7 June 1984, Detroit, USA (SME Technical Paper MS84-362). Reproduced by permission of the author and the Society of Manufacturing Engineers.

Efforts to automatically assemble existing products not designed for robotic assembly are usually costly and difficult to implement. Design guidelines should be incorporated into the design of a product so that a robotic or automated assembly and test process is possible. The following subjects are discussed as they relate to product design: layered assembly, mechanical design, feeding small parts, fasteners, flexible parts, labels, packaging, automatic testing of electrical devices, economic justification for design changes and design for automation misconceptions. Examples of desirable and undesirable design characteristics are presented along with general guidelines for design for robotic and automatic assembly and test.

Efforts to automatically assemble existing products not designed for robotic assembly are usually costly and difficult to implement. The manufacturing process must be considered when the product is designed, not after the design has been completed.

Often assembly of a product would have been easy to automate for one feature making it impossible or economically unjustifiable to automate. With just a few minor design changes, and usually with no effect on the hardware cost of the product, the product could have been automatically assembled.

Layered assembly

One of the first considerations of robotic assembly is to avoid lifting and rotating the assembly. This results in complicated fixtures and grippers, more degrees of freedom required in the robot, and increased cycle times. A reduction in handling will simplify the assembly process and can be accomplished with a layered assembly design.

A layered design enables the parts of the assembly to be stacked without any rotation or regrasping of the parts. Each part is vertically placed on the

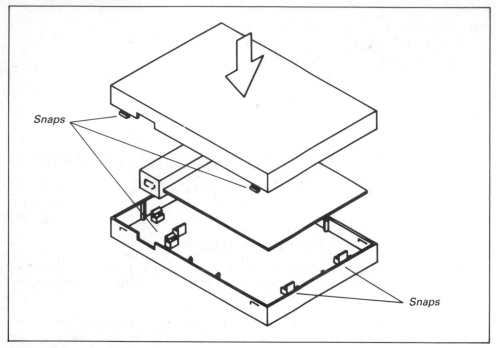

Snaps

Snaps

Fig. 1   Example of a layered assembly

one below it, thus eliminating the need to lift or rotate the unit during assembly. This avoids the use of turnover fixtures and rotary tables.

Typically, the bottom cover is placed on the assembly fixture first, followed by the parts, layer after layer. If fasteners are required because the parts do not snap together, they should also be orientated vertically. The final step is to place the top cover on the bottom cover as illustrated in Fig. 1.

Having a product designed in this manner reduces assembly to a series of pick-and-place operations, thereby requiring a less sophisticated robot. This results in manufacturing cost savings and increases the likelihood of financially justifying robotic assembly.

Mechanical design

The major mechanical design concepts to be considered when designing a product are now discussed.

Compliance

One of the most important concepts to consider in the design of a product is the accommodation of uncertainty, or compliance. Since all parts are only dimensionally stable within certain tolerances, this uncertainty must be accommodated in the design. This also applies to robots and other automatic equipment. In spite of their accuracy and repeatability, a robot can never move to exactly the point in space it was programmed, nor can it return to exactly the same spot. Therefore, these variances in parts and equipment movement must be considered in the design of a product.

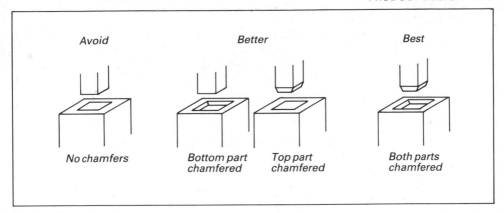

Avoid   Better   Best

No chamfers   Bottom part   Top part   Both parts
                   chamfered   chamfered   chamfered

Fig. 2   Compliance

One of the easiest ways to design compliance into the parts of a product is to allow the parts to fit together even when misaligned. Fig. 2 shows how this can be accomplished through the use of chamfers, ramps, and lead-ins. The fit between parts should be a positive clearance fit; interference fits should be avoided.

Self-alignment of parts

Parts should fit together only one way and not require a secondary operation for alignment. Guide pins, D-shaped holes, and notched slots are examples of ways that parts can be assembled in only one configuration.

Another example of self-alignment of parts is that of nesting. A nest is a hole, cutout, or trough designed to secure a part in a particular orientation. The concept of nesting works well with the concept of layered assembly since nesting allows a part to be placed and orientated before it is secured (see Fig. 3).

Hidden features

Parts with hidden features such as holes, slots, pins, and so forth that must be orientated with respect to these features, should have corresponding external features to help orientate the part. These external features, as shown in Fig. 4, need not be functional other than to help define the orientation of the part.

Symmetry

When possible, parts should be symmetrical so they may be fed and assembled in more than one direction. Designing a symmetrical part will reduce the need for sensors to detect features and reduce handling. Fig. 5 shows three parts that could be redesigned to take advantage of the concept of symmetry.

If a part must be asymmetrical, it should be clearly asymmetrical. If it is close enough to being symmetrical to be difficult to determine its orientation, the part should be treated like a part with hidden features and

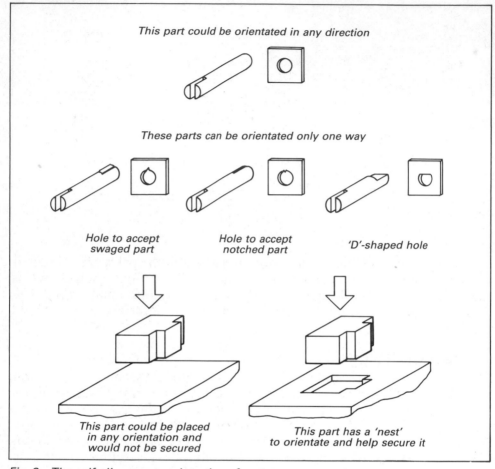

Fig. 3   The self-alignment and nesting of parts

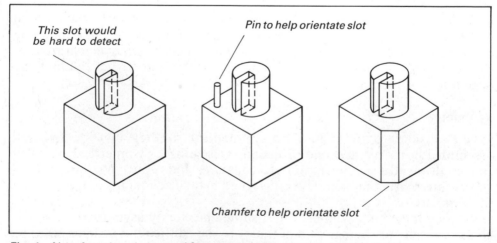

Fig. 4   Non-functional external features to help orientate hidden features

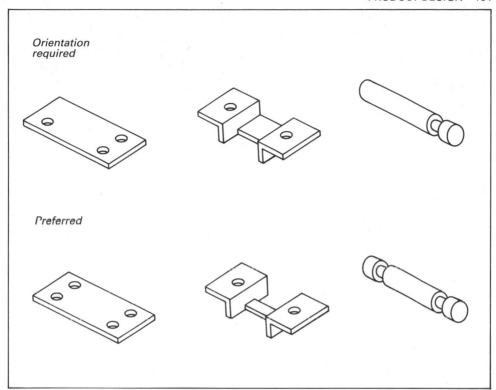

Fig. 5   Examples of symmetry

have a nonfunctional external feature to increase its asymmetry. This will help to eliminate part feeding jams and misalignment of parts during assembly.

Parts designed for grippers

Parts should be designed to be compatible with gripping. Parts with large, flat and smooth top surfaces are the easiest for vacuum or magnetic pick-up. If a part cannot be picked using vacuum or magnetics, it will have to be picked using a mechanical gripper, and should have appropriate holes, slots, or tabs to aid grasping.

Feeding small parts

Many small parts can be fed using a vibratory bowl. It is very important to ensure that these parts can be bowl fed and will not jam or tangle in the bowl or slide.

Parts tangling

Parts should be designed so they cannot interconnect or tangle, as illustrated in Fig. 6. If springs must be used, use closed loop springs with the diameter of the wire greater than the spacing between coils. An alternative that would be easier to handle than a coil spring would be a leaf or flat spring.

This part can tangle easily

The same part redesigned, will not tangle

Parts that interconnect
will not feed

Springs with open loops
will tangle

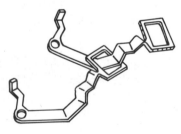
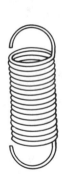
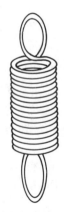

A fillet will keep the
parts from interconnecting

Springs with closed loops
will not

Fig. 6   Small parts tangling

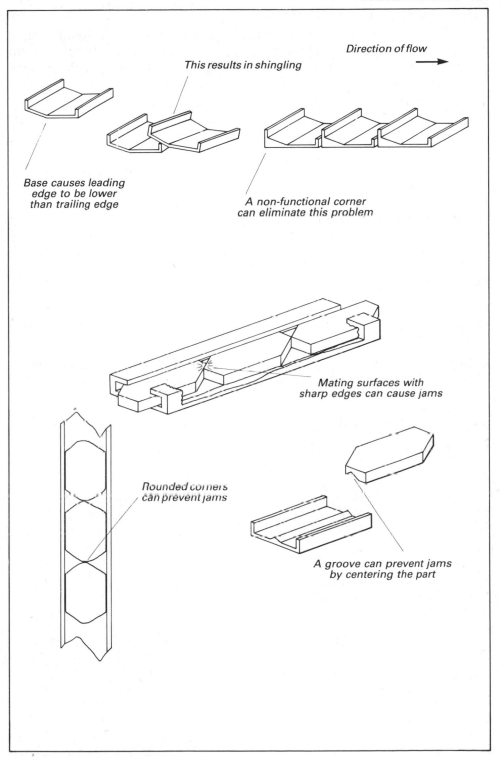

This results in shingling

Direction of flow

Base causes leading
edge to be lower
than trailing edge

A non-functional corner
can eliminate this problem

Mating surfaces with
sharp edges can cause jams

Rounded corners
can prevent jams

A groove can prevent jams
by centering the part

Fig. 7   Methods to avoid jams

Parts stability to prevent jamming

To enable small parts to be fed down a slide, they must be designed so they do not jam or shingle in the track. Examples of jamming and shingling are shown in Fig. 7. The following is a list of design guidelines to help prevent jamming when feeding small parts:

- Sharp edges should be avoided to prevent the parts from catching on the slide.
- The mating surfaces of the parts should be perpendicular to the direction of travel and be as large as possible.
- The bases of the parts should be flat to prevent the parts from riding on top of one another.
- To increase the stability of the part during feeding and assembly, it should have the lowest centre of gravity possible.
- A slot or rib in the part, centred on the bottom of the part parallel to the direction of travel, will help stabilise the part and keep it from jamming.

Fasteners

Parts can be joined or fastened together without the use of fasteners such as screws, bolts, clips, or rivets if they are designed so they may be snapped together. The following are facts to consider:

- If possible, have parts joined or connected without using fasteners by having them designed so they may be snapped or interlocked together (Fig. 8).

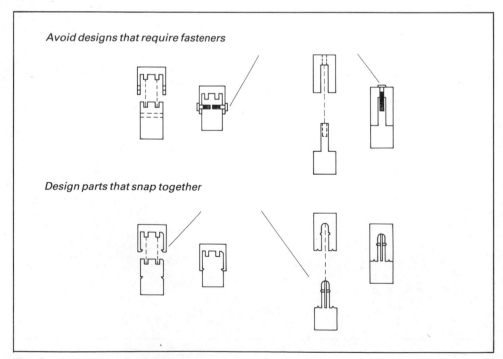

Fig. 8   Substitutes for fasteners

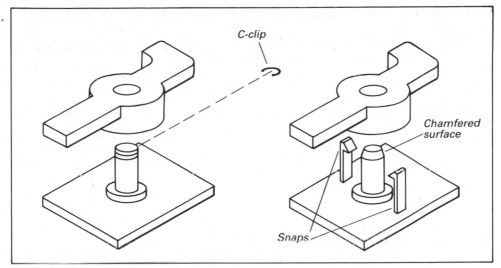

Fig. 9   Joining moving parts without fasteners

- Moving parts may also be snapped together as shown in Fig. 9.
- If the use of fasteners cannot be avoided, use fasteners that can be easily fed automatically. Fasteners with a length to diameter ratio of at least 1.5 to 1 are most easily fed.
- If a washer is required, have it captured by the screw or bolt. It should not be a free part.
- Use the same fastener throughout the entire assembly. Avoid using different sizes and types of fasteners.
- Fasteners orientated vertically are the most accessible to automatic equipment.

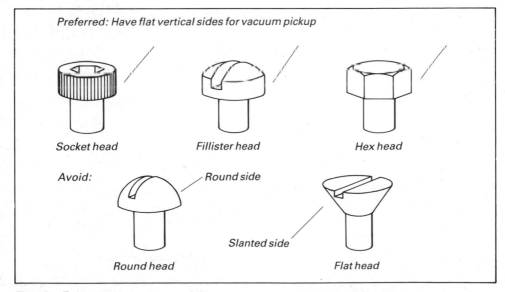

Fig. 10   Fasteners

- Always allow adequate clearance around the fastener for an automatic tool.
- Use fasteners with flat sides and tops so they may be more easily picked up magnetically or by using vacuum (see Fig. 10).

Flexible parts

A flexible part is more difficult than a rigid part to assemble because the position of the part is variable. All parts that are to be assembled robotically should be rigid and stiff. Even parts like covers and frames can bend under their own weight and should be strengthened, if required, by using a thicker or stronger material, or by using reinforcing ribs.

The most difficult flexible parts for robotic assembly are belts, wires and cables.

Wires and cables

Wires and cables are difficult to locate and orientate due to their flexibility. However, there are alternatives to using wires and cables to connect various components in an assembly. The components can be plugged directly together, thus eliminating the connecting wire or cable. For example, the electrical connection between a circuit board and a power supply can be made by plugging the two together instead of having them joined by a cable. A switch or indicator can be plugged or soldered directly to a circuit board to form a connection without the use of wires.

Parts can also be integrated to eliminate connecting cables. Instead of having multiple logic cards, locate all the logic on one large card so that it may be plugged directly into the power supply.

If the two components to be connected cannot be plugged directly together, they still do not have to be connected by a wire. A slave circuit board with only continuity circuitry can be used instead of a set of wires to make the electrical connection. The slave circuit board would plug into both components and provide the electrical connection between them (Fig. 11).

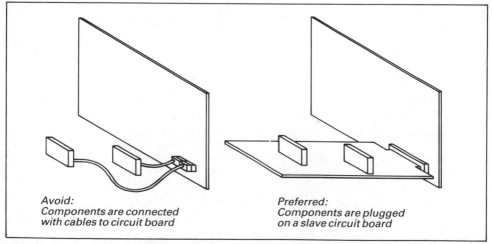

Avoid:
Components are connected
with cables to circuit board

Preferred:
Components are plugged
on a slave circuit board

Fig. 11   Example of a slave circuit board

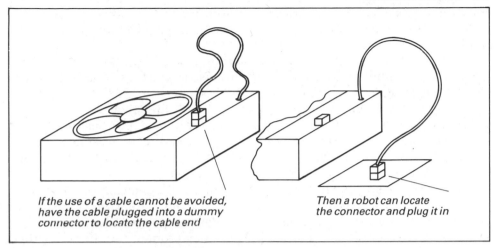

If the use of a cable cannot be avoided, have the cable plugged into a dummy connector to locate the cable end

Then a robot can locate the connector and plug it in

Fig. 12   Example of a secured cable

If the use of a cable attached to a part cannot be avoided, the cable end must be in a known location and orientation for robotic picking. This can be accomplished by having the cable end plugged or screwed into the part when the part is manufactured. The robot can then locate the cable end, unplug or unscrew the cable, and connect it to the proper location in the subassembly as illustrated in Fig. 12.

Belts

Belts have the same problem as wires and cables.They are very difficult to feed, locate, pick up, and place on a pulley or wheel. If possible, belts should be substituted with gears or drive shafts for torque transmission.

Labels

Installing a label on an assembly can be more difficult and labour intensive than installing a part. Also, on many simple assemblies there may be more labels than parts, which means that a significant portion of the total labour will be for label application. Consequently, a significant labour reduction can be realised if labels can be applied automatically.

The first step to take when trying to influence label design is to determine if the label is needed. Try to eliminate unnecessary labels or combine them with other labels to reduce the number that must be applied. For example, an Underwriters Laboratories safety label could be combined with a power requirement, serial number or machine type label.

The label application problem can be eliminated by having the label moulded, painted, or stamped when the part is manufactured. This is the most desirable technique and should be pursued if possible.

If the labels cannot be incorporated into the part itself, the possibility of having the labels installed by a vendor should be investigated. For example, if the assembly has moulded covers that must have a logo or label applied, the vendor may be able to apply it with little additional effort when it leaves the moulding machine. A sourcing decision should be made comparing the additional vendor charge to apply the labels against the in-house cost of label application.

As mentioned previously, a consideration of automation is to reduce lifting and rotating the assembly, and this also applies to the application of labels. Regardless of the application method used (manual or automatic) handling of the assembly can be minimised if all the labels can be located together. They should all be on the same surface or side of the assembly, and in the same area on that surface or side. If the labels are in the same area, it is much easier to justify automatic label application equipment since only one label applicator may be needed.

Different types of label application equipment are available. The most expensive yet most adaptable are laser printers. Depending on the texture of the surface, they can burn a label into plastics, paper, and even certain metals. Since laser systems are expensive, they probably could be justified only on high-volume assemblies.

If a laser cannot be economically justified or is not technically feasible, self-adhesive paper labels on rolls can be applied automatically. These need a smooth flat area for application and may be limited to what may be printed on them.

Self-adhesive paper labels on sheets should be avoided since they are difficult to feed and remove from their backing automatically. Labels that require glue or solvents for retention should also be avoided.

The following items have been arranged in the priority in which they should be considered:

- Determine if the label is really required.
- Have the label designed into the part, e.g. moulded in, stamped in or painted on.
- If economically justified, have the vendor who manufactures the part install the label.
- If the label must be installed in-house, have all labels on the same area of the assembly.
- If the labels cannot be in the same area, have them on the same side or surface of the assembly.
- If it is economically justifiable and feasible, use a laser printer to burn-in labels.
- Use self-adhesive labels on a roll.

Packaging

Packaging design for the parts and finished assembly, as well as the design of the product to withstand shipping, are now discussed.

Packaging for parts presentation

Usually the greatest effort associated with robotic assembly is not with the actual assembly process, but with the handling and presentation of the parts and subassemblies to the robot. If the packaging does not present the parts in a known and repeatable location, then they must be manually fed, resulting in double handling. If the parts must be manually fed to the robot, then the line stocker may just as well perform the assembly operation. All manual effort must be eliminated to have a truly automatic process.

Ideally, the line stocking process should be eliminated by having the robot pick its own parts directly from the box or carton. However, this requires packaging designed for precise parts location. Trays, compartments, magazines, or partitions can be used for locating the parts in known positions within the box. The part should be securely located within its compartment or partition and have its most easily 'pickable' side facing the approach angle of the robot. The box itself should be dimensionally stable to allow it to be placed in a known and repeatable location.

The boxes the parts come in as well as the boxes the final unit is to be shipped in should have lids instead of flaps to facilitate automatic removal. Otherwise, complicated flap opening devices will have to be used.

There may be a benefit to ship the final product in the same container one of the parts came in. For example, reuse the box the covers of the product came in since the final product should be the same size as its covers. In other situations it may be better to pick parts directly from a power pack on a skid instead of individual boxes. The most economical method that precisely locates the parts should be chosen.

Product design to simplify packaging

An area not often considered is how the product is to be designed to withstand the various shocks, loads, and vibrations it will experience during shipping without requiring excessive or unusual packaging. The simpler the packaging of the product, the easier it will be to have a robotic packaging station that uses standard cartons and pallets.

Whenever possible avoid the need for the support of internal parts in the product with packaging material. Also, avoid openings in the product in which packaging material may enter and contaminate the product. If possible, have the carton designed so the unit may be placed straight down in the shipping container, with the packing material already in place.

Automatic testing of electrical devices

The product should be designed so that it may not only be assembled automatically, but also tested automatically. Design guidelines to be followed so that an electrical device may be tested automatically are:

- There should be no internal test points in the product that must be probed at final test time. Internal points that must be tested should have parallel external test points.
- All test points, connectors, and switches should be readily accessible and preferably on the same surface of the product, as shown in Fig. 13.
- Locating pins or holes should be on the test point or connector surface to help guide and align the test probe or fixture.
- Instead of having the power cord wired directly into the power supply of the product, use a pluggable line cord. The pluggable connector on the power supply can be easily probed to supply power during test.

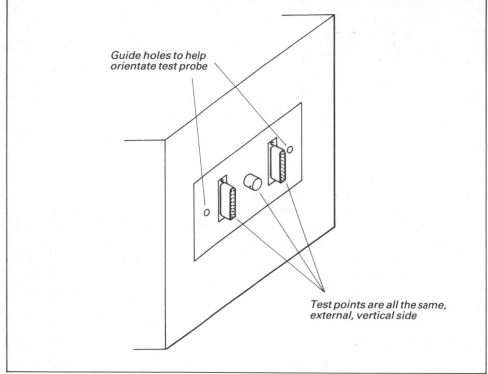

Guide holes to help orientate test probe

Test points are all the same, external, vertical side

Fig. 13   Example of external test points

- Use only standard connector types that have lead-in or chamfered surfaces to help guide the test probe.
- If adjustments must be made at final test time, use knobs and screws that are readily accessible and easily manipulated.
- The connector mounting surface on the product must be able to withstand the force required to plug and unplug all connectors simultaneously.
- To ensure the accuracy of the position of the test points when the unit is in the test station, locate their positions in reference to the base or a side of the product as accurately as possible.
- Do not have any wires or flexible parts that must be probed and tested.
- Avoid designing machines requiring warm-up and burn-in chambers since these chambers will complicate the test process.
- Design machines that can be tested independently without being connected to other machines. For example, design a video display that does not have to be connected to a logic unit ot keyboard to be tested.
- Provide common test connector points within a product family.

Economic justification

Changes made to a product design to allow for automation may result in an increase in hardware cost. When the cost of hardware is increased as a result of a design for automation change, this change should be economically justified before it is incorporated into the design of the assembly.

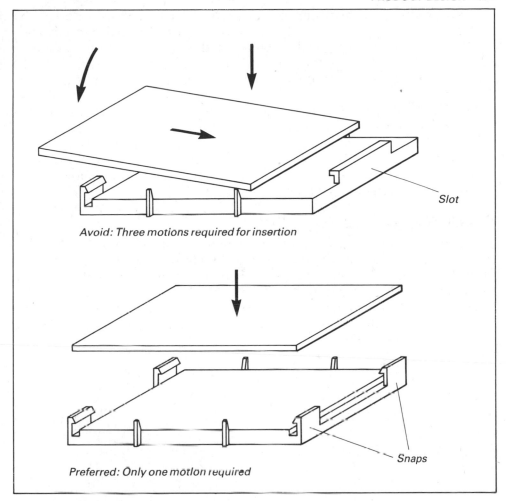

Avoid: Three motions required for insertion

Slot

Preferred: Only one motion required

Snaps

Fig 14   A design improvement

The additional hardware cost of the product will have to be compared with the labour saved and the capital investment required by an automated process. The depreciation and investment tax credit of the automation equipment as well as any operational expenses should be included in the economic justification. With all the savings and costs known, a decision can then be made regarding the impact of the design change.

Design misconceptions

As in any other field, a little bit of of knowledge can sometimes be counterproductive. For example, it is desirable to design a product without fasteners and to have it snap together, but the resulting design may require such extensive repositioning and rotating of parts to make robotic assembly almost impossible. In this case, if the product was designed with screws that were accessible, it would be much less expensive to use an automatic screwdriver than to use a robot that can handle the complex motions.

Some products can be assembled automatically in their current design, but still can be improved. A product could be snapped together easily by a human operator, but still involve complicated motions for a robot. Without changing the function of the product, or increasing its cost, it could be designed for the simple automation that a layered assembly design allows (see Fig. 14).

Do not blindly adhere to a particular automation guideline since another guideline may be more applicable. As mentioned in the previous paragraph, a guideline such as eliminating fasteners in a product may not always be the best one to implement. It is very important to consider the proposed manufacturing process when designing a product for automation and have the product designed accordingly.

Concluding remarks

A robotic or automatic process has the potential of significantly lowering the manufacturing cost of a product and should be installed if possible. However, it is much more difficult and expensive to install a robotic assembly and test process for a product that was not designed for automation. With the aid of these guidelines, a product can be designed for automation from the beginning and fully realise the benefits of a robotic manufacturing process.

CONSIDERATIONS FOR ASSEMBLY ORIENTATED PRODUCT DESIGN

R. D. Schraft and R. Bässler
IPA, West Germany

First presented at the 5th International Conference on Assembly Automation, 22-24 May 1984, Paris. Reproduced by permission of the authors and IFS (Conferences) Ltd.

A study to investigate the state-of-the-art of assembly automation showed some interesting results concerning assembly orientated product design. One major point was the question: what engineering know-how (guidelines, catalogues) for an assembly orientated product design can be found in industry and to what degree is it implemented in current product developments? A prior condition for the realisation of assembly orientated product design is good cooperation between engineering and design departments. Furthermore, the design engineer must have access to user-orientated models for the application of this know-how. Based on these facts and on the results of the study, the possibilities for a systematic procedure to apply the existing know-how in assembly orientated product design are presented.

In recent years, rules and guidelines for assembly orientated product design have been proposed, with the aim of achieving an assembly procedure requiring the least possible amount of time, assembly devices, space and personnel[1]. These efforts are influenced and often supported by an increasing amount of new joining technologies, making simpler, more cost efficient and often more automated assembly possible.

Due to rising assembly costs, demands on the design engineer for assembly orientated product design are ever increasing; even though today over 15 criteria for manufacturing machining orientated design must already be considered (function suitability, serviceability, NC-orientated element design, etc.). The reason for the necessity of assembly orientated product design is primarily because rationalisation attempts made exclusively in the assembly process without regard to other manufacturing steps, only enable a relatively small decrease in the overall costs of the manufacture of a product[2,3], in spite of extensive efforts.

The problems surrounding assembly orientated product design were examined in a study involving 3500 companies[4]. Among numerous other

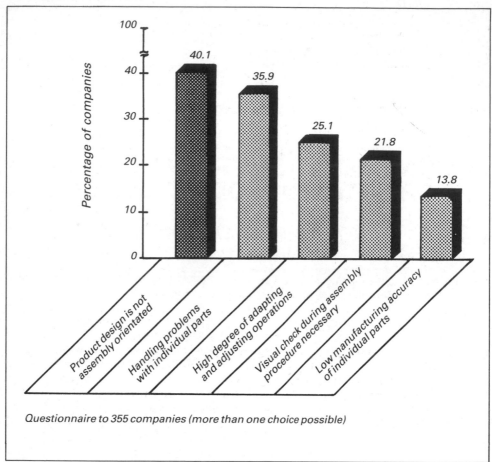

Fig. 1   Obstacles to automation in product design

problems, this study examined and analysed the obstacles for product design, possibilities of reducing production time, the available and applied means to achieve an assembly orientated product design, as well as relevant 'technology' forecasts.

Several of the most important results are shown here. In many companies, insufficient consideration was given in the product design stage to the requirements for assembly orientated design. Fig. 1 shows the most frequently encountered obstacles to assembly automation with regard to product design. The two most important being that the product design is generally not assembly orientated and that most parts cannot be handled automatically without problems.

The correlations shown in Fig. 1 were also confirmed in a Delphi forecast as a part of the study; 76 experts in the assembly field participated (Fig. 2). The majority of the experts questioned estimated that by 1987 assembly orientated product design would have the same importance as NC-orientated design today. As well as enabling or simplifying assembly automation, positive effects through assembly orientated product design are

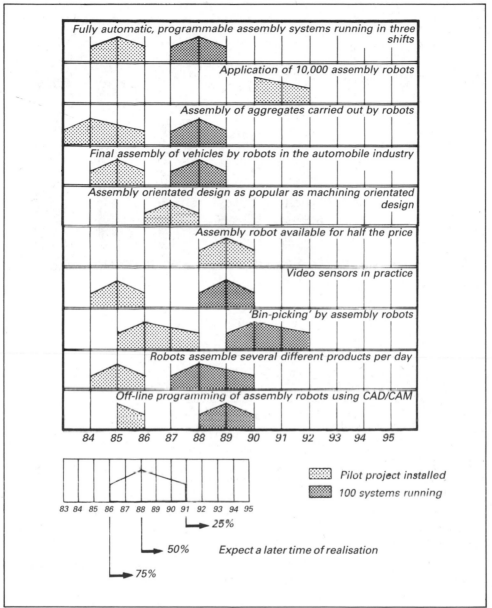

Fig. 2   Delphi-forecast: industrial robot-application in the assembly field (1983)

expected, among other things, on the assembly time allowed. Fig. 3 shows the percentage of companies expecting a reduction in assembly time through individual assembly orientated product design measures. Above all automation-orientated product design and alterations to existing product concepts are expected to positively influence the assembly time allowed. There is also a high percentage of companies expecting a reduction in assembly time through a decrease in the number of parts or through simplified joining connections

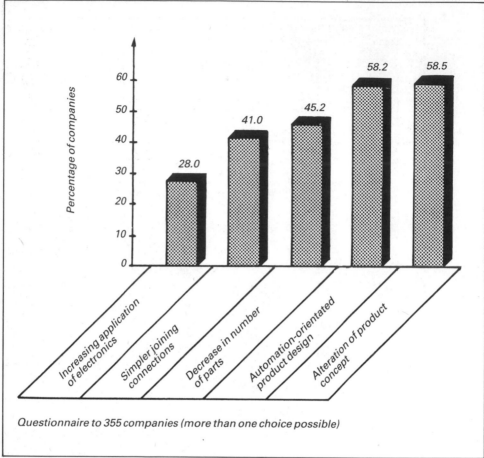

Fig. 3   Percentage of companies expecting to reduce the assembly time allowed through assembly orientated product design measures

However, it is often not possible to fulfil these high expectations since the know-how in this field is not sufficiently available or because no systematic procedures are presented, to convert assembly orientated product design measures into the design process in industry[5].

Design strategies and measures

Assembly orientated product design leads to a decrease in assembly automation costs, the possibility of automation, or reduced manual assembly effort[6]. Corresponding to the level of planning, there are various strategies for an assembly orientated product design, which in turn result in various degrees of rationalisation potential (Fig. 4)[7]. Measures concerning the complete product structure lead to the greatest savings. However, they may only be realised efficiently in long-term projects in the new design of products or product types, since alteration expenses to an existing product structure would exceed the savings expected.

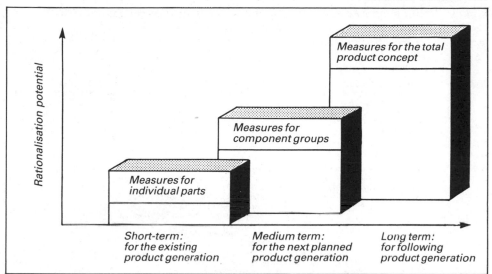

Fig. 4 Strategies necessary for an assembly orientated product design

It is possible to carry out design measures for component groups in existing products at relatively low expense; but it is important that interfacing points to other component groups and parts are not left unaltered. The rationalisation potential of these measures, however, is lower than that of the product structure measures. One example of altering

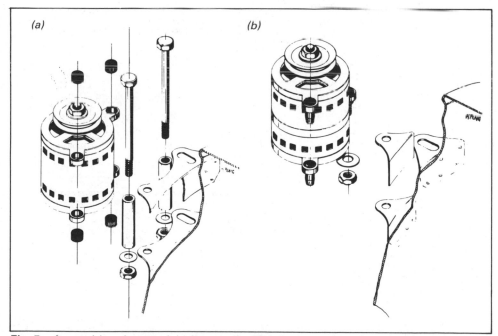

Fig. 5 Assembly orientated design applied for the mounting of an electric motor onto a washing machine tub: (a) assembly of 13 individual parts and (b) assembly of 3 individual parts

Table 1   New joining-related technologies for assembly orientated product design

Technology	Functional principle	Results	Examples
Outsert technique	Plastic parts with mechanical functions are injection moulded on a plate in a single pass and without any finishing operation	Assembly of the functional parts on a plate doesn't take place. No joining parts (e.g. screws) are necessary. High degree of dimensional accuracy of the parts on the plate. Disassembly is usually not possible	Switch clocks Switch boards Clockwork-chassis for video recorders Plate of a printing saddle for a matrix printer
Insert technique	At the injection moulding machine the functional parts (mostly of metal) are inserted in the tool-form and moulded	Assembly of the functional parts on the basic part doesn't take place. No joining parts are necessary. Assembling operations are transferred in the part manufacturing process, which can be automised without any problem. Disassembly is not possible.	Contact springs for car rear lights
Bonding by adhesives	Joining of parts with adhesive. Because of the great improvements in the field of the bonding technology it is possible to bond the different materials with a high strength	A great number of joining parts (screws, rivets, etc.) are not necessary. Assembling operations are more simplified and automaticable. This method necessitates a higher degree of quality inspection. Sometimes curing time in an obstacle for automation.	Bonding in of windscreens in cars Attachment of car body parts (e.g. pads) Fixing of caps on car motors Bonding of a shaft-hub-joining (instead of shrinking)
Snap fitting	The elastic plastic parts are moulded and formed in a way, that the parts are locking in each other by a straight joining movement.	Joining parts are not necessary. The assembling operations are therefore very simple and can often be automised. Generally higher tooling-costs. Depending on the design, disassembly is possible.	Wall clamps for pipes Many applications in the household appliance industry
Thick-film hybrids	Printed metalliferous paste is burnt into a carrier material to form resistors, conductor tracks and insulating points, resulting in real printed circuitry; these may later be completed by placement of components.	Miniaturisation of electric, very reliable circuits showing high degree of temperature and chemical stability. Very accurate structures and values may be produced. Joining parts and the assembly of complicated flabby parts are not necessary. Disassembly or reassembly are not possible.	Numerous applications in professional communication techniques, industrial electronics and consumer electronics
Thin-film hybrids	Application of conductor tracks and resistors out of pure metal (thickness of layers in vicinity of light wave length) onto a substrate (carrier material). Metal layers are implemented in two ways: condensation of vapourised metal and high speed blasting of pulverised metal		

component groups is the mounting of an electric motor onto a washing machine tub (Fig. 5). Of the previous 13 parts to be assembled in several joining directions, only three parts remain to be assembled in one single joining direction.

The lowest rationalisation potential may be found in carrying out measures on individual parts. Such measures, however, may often be carried out at a low alteration expense on existing products, but in some cases greatly influence the overall cost of automatic assembly of the product. Recent developments in 'joining' technology is the best example of such measures.

Assembly orientated joining techniques

The representative questionnaire showed that 41% of companies expected the assembly time to be reduced by simplifying joining connections.

Conventional joining techniques, such as screwing, fulfil the many demands placed on a joint, such as reliability and serviceability; however, this type of joining is expensive both for parts manufacturing and assembly automation[8].

Fig. 6   Outsert technique: drive for tuner mechanism

Fig. 7   Assembly orientated product design of car rear lights

The best possibilities for simplifying joining operations can be found in the recent developments in the field of plastics. Joining operations, such as the insert and outsert techniques described in Table 1, require relatively expensive parts production, but this is compensated by the simplified assembly procedure. The possibilities of such techniques and typical examples are shown in Figs. 6 and 7. In the case of the tuner mechanism, the plastic gears are moulded directly onto the metal plate. This technique eliminates the complete assembly of the gears, including the joining parts (about 28 individual parts) necessary in conventional joining operations.

The use of plastics provides a further advantage, i.e. the possibility of forming complicated shapes and the integration of parts[9].

Systematic design procedure

Documentation

To enable the design engineer to implement measures for assembly orientated product design, he must have access to user-orientated aids. It is therefore necessary to carry out intensive investigations into joining techniques.

First, a systematic listing, classification and description of several assembly orientated joining techniques should be compiled. In addition to outlining the technical procedures, salient features, advantages and disadvantages, specific application requirements, boundary conditions, etc. should also be described.

Table 2 Technical/economic comparison of joining techniques [10]

Joining techniques	*Strength*	*Assembly cost*	*Design*	*Reliability*	*Visual inspection*	*Ease of maintenance*	*Alignment accuracy*	*Flexible alignment*	*Small parts*	*Large parts*
	Main features									
Screwing	1	3	3	1	1	1	2	2	3	1
Resistance welding	1	1	2	3	3	3	3	1	1	1
Arc welding	1	2	2	1	2	2	3	1	3	1
Hard soldering	1	3	1	1	2	3	3	1	1	1
Rivetting	1	2	2	1	1	3	1	3	3	1
Notching	3	1	3	2	1	3	1	3	2	3
Overlapping	2	1	1	1	1	3	3	2	1	3
Bonding by adhesives	3	2	1	2	3	3	3	1	1	2
Special joining elements	2	3	3	1	1	2	2	2	3	1

1, most favourable; 2, favourable; 3, good

In compiling such 'catalogues' of information, the relative costs of joining techniques should be compared from both practical and economic viewpoints. In addition to the costs of part production and the assembly of the joint, features such as joint reliability and strength must also be considered (Table 2). In this context it is clear that very economical joining procedures for assembly have been developed (e.g. overlapping, notching), whereas screwing, for instance, represents a relatively expensive technique.

Rules and guidelines for assembly orientated product design, valid for all industrial applications, must also be set-up and summarised in appropriate surveys; thus enabling the design engineer to apply these means in a user-orientated manner in practice. Such rules/guidelines should be presented with an overall product concept in mind (taking into account organisational sequences, flow of materials, etc.), using descriptions and examples where possible. Subdivision of information is most useful since it enables the rules to be put together in a relatively compact form. More detailed information, such as actual examples, may be handled in an 'extended information' section.

Due to the different demands and special peripheral conditions in the individual product spectrum, the design guidelines for each industrial area should be put together separately. General valid guidelines for assembly orientated product design do exist, but the more complex and specific the features of the product, the less valid such guidelines become.

Computer data collection

Application of the guidelines in industrial situations often fails due to the large amount of information which is not systematically classified. The design engineer must collect information and design guidelines for his particular problem from various different sources. The individual pieces of information often have different objectives, making direct application to the actual design impossible.

An efficient work procedure for the application of the design criteria to product construction is impossible, except when using computer systems. If it is possible for the design engineer to obtain the necessary information quickly and precisely he is able to incorporate appropriate guidelines into his design. A computer based system for assembly orientated product design must consist of the following components:

- Computer data collection of joining techniques as well as rules for assembly orientated product design, the design of component groups and individual parts.
- A principle for determining the design efficiency of a product dependent on the design stage.
- A procedure for systematic computer based design using guidelines and algorithms for the determination of the design efficiency of a product.

Such a computer based procedure makes it possible for the design engineer to check if the design is assembly orientated in each respective design stage and, if necessary, to alter the design, successively increasing its design efficiency.

Computer based systems such as those described above have not yet been developed. The necessary operations are therefore described and starting points for solutions shown.

Computer data collection for joining techniques

The constant changes and developments in joining techniques make data collection in this field difficult to compile. Data must be set up in an expandable form, preferably in the form of a design catalogue. Subdivision of the joining techniques into branches or procedures enables the product specific boundary conditions of individual joining technologies to be considered.

Data collection or rules must be subdivided into criteria of various different levels (Fig. 8). Due to product specific differentiation criteria in individual areas, the highest differentiation level was chosen to be the subdivision of these branches. The next level of differentiation was chosen to be the dependence of individual guidelines on devices. The dependence of guidelines on devices relates, for example, to a necessity resulting from particular features of ordering and handling units. The third level of the subdivision refers to the general validity of guidelines. The guidelines which are the same in each branch, i.e. the generally valid guidelines, are listed separately. The next level distinguishes between the individual design stages, depending on the design progress. The design stages are shown in the main part with concrete guidelines for assembly orientated product design and further examples are described in the extended information section. The subdivision of design guidelines presented here are also compiled in an expandable form, enabling them to be constantly expanded and supplemented. The amount of data on design guidelines makes it advisable to

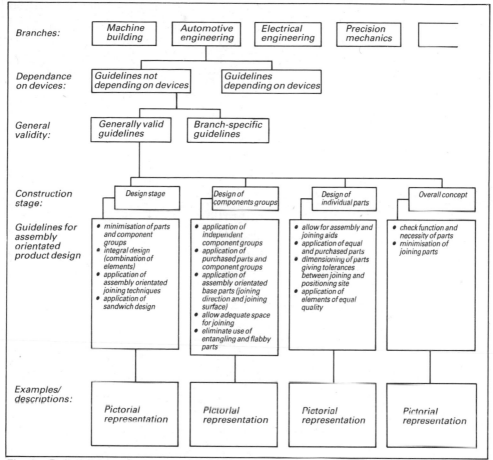

Fig. 8   Subdivision of a computerised file

store this in a computer data base and furthermore to harmonise it with a computer-based system for assembly orientated product design.

Determining the design efficiency of a product

The aim of the computer-based system for assembly orientated product design is to achieve an assembly orientated product structure by successive improvements in small areas of the design stage. Prerequisite for an improvement is a quantitative statement of the respective level of design efficiency of a product or components acquired. In order for such statements to be made, parameters characterising design efficiency must be determined. Algorithms for determining design efficiency must be set up and appropriate characteristic factors for these must be determined.

The characteristic factors regarding the design efficiency of a product must, amongst other things, consider the influence of costs, part numbers, batch sizes and variant numbers. A decision, based on these factors, must be made as to whether a further increase in design efficiency is possible by assembly orientated product design measures.

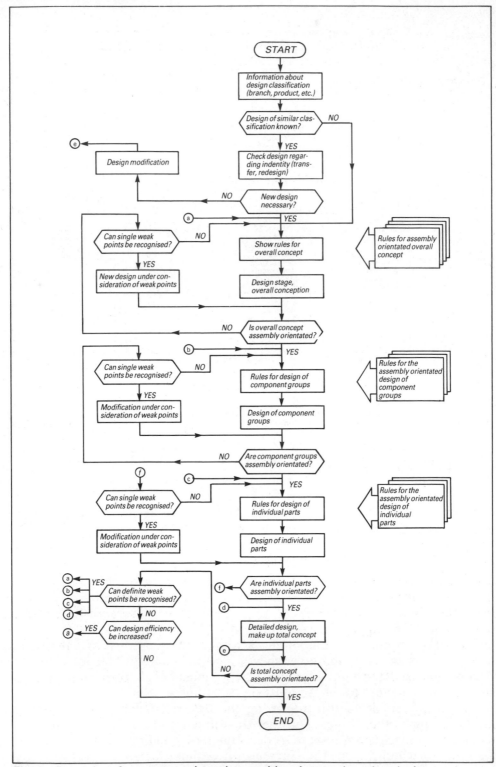

Fig. 9   Procedure for computer-based assembly orientated product design

Procedure for a computer-based assembly orientated product design

The procedure for a computer-based assembly orientated product design is based on data collection on joining techniques and their design guidelines, and on the principle for determining the design efficiency of a product. This enables the respective industrial-area specific information for the individual design stages to be prepared (Fig. 9). The individual design stages are carried out under consideration of these guidelines. At the end of each stage it is tested if the design is assembly orientated. Independent of this result, the design will be altered or redesigned accordingly. After all design stages have passed through this iterative process, the design efficiency of the overall design is tested and the possibility of an increase in design efficiency examined.

The procedure for computer-based assembly orientated product design is based on CAD. The respective design guidelines may be 'called' onto the screen using a so-called menu-technique. The necessary design characterisation information, followed by information for determining design efficiency, are entered in dialogue communication. Several data necessary for characterising the design may be taken over directly from the computer-based procedures and complement each other, forming a useful unit.

Products which are not designed assembly orientated are frequently the reason for uneconomical assembly systems. Furthermore, systematic procedures for the application of assembly orientated product design effect a more cost-efficient production, especially in assembly, but also in part production. Companies introducing computer-based procedures can expect immediate economic advantages.

References

[1] Boothroyd, G. and Dewhurst, P. 1983. *Design for Assembly – Handbook*. University of Massachusetts, Amherst, USA.
[2] Andreasen, M. M., Kähler, S. and Lund, T. 1983. *Design for Assembly*. IFS (Publications) Ltd, Bedford, UK.
[3] Lund, T. and Kähler, S. 1983. *Design for Assembly*. Proc. 4th Int. Conf. on Assembly Automation, pp.333-349. IFS (Publications) Ltd, Bedford, UK.
[4] *A Study for the Ministry of Research and Technology*, 1984. IPA, IAO, GfAH and others.
[5] Swift, K. G. 1983. A computer based design consultation system. *Assembly Automation*, 3 (3): 151-154.
[6] Abele, E., Walther, J. and Bässler, R. 1983. *Montageautomatisierung mit Industrueroboter*. Flexible Automation, Mikrocentrum Eindhoven/Enschede.
[7] Köhne, F. 1983. *Praxis der Montageautomatisierung am Beispiel der Automobilindustrie*. Tagung Praxis der Montageautomatisierung, Fellbach, VDI-Berichte 479, pp. 1-13.
[8] Lotter, B. 1983. *Montageerweiterte ABC-Analyse*. Deutscher Montagekongress, Munich, pp. 53-71.
[9] Kohn, H. 1983. Fertigung von Kunststoffkraftstofbehältern (KKB) mit hohem Automatisierungsgrad. *Kunststoffe im Automobilbau* pp.157-191. VDI-Verlag, Düsseldorf.

[10] Ehrenspiel, K. 1983. Wahl der kostengünstigsten verbindung. *Spektrum der Verbindungstechnik – Auswählen der besten Verbindungen mit neuen Konstreuktionskatalogen*, Köln. *VDI-Berichte* 493, pp. 175-182.

4

Programming Systems

A selection of state-of-the-art and forward looking contributions consider the present variety, capabilities and relative merits of different robot programming systems. The present variety of robot programming languages is examined and the motivation for ongoing development in this important field is discussed.

ROBOT LANGUAGES IN THE EIGHTIES

G. Gini
Politecnico di Milano, Italy
and
M. Gini
University of Minnesota, USA

First presented at the 1st Robotics Europe Conference, 27-28 June 1984, Brussels.
Reproduced by permission of the authors and Springer-Verlag.

The scenario of general purpose programming systems is rapidly changing; so what are the consequences for robot programming? After introducing the peculiar aspects of robot programming some examples of general purpose languages applied to robots and languages specifically designed for robotics are discussed. Criteria for making a choice between the two approaches should take into account the present state-of-the-art. The need of a strong integration between different components (robots, vision systems, and other automation equipment) could support the first approach. The solution of robot and robot users' problems has, until now, supported the second approach, as indicated by the choices of European robot manufacturers. Nevertheless, the expression of actions taken by different intelligent agents, as robots can be defined in the future, will require absolutely new linguistic media. It is hoped that robot programming in the 'eighties' can develop experiences useful for the assessment in the field.

Today robot applications are generally carried out in an integrated industrial setting, where robots and other equipment manipulate and sense parts to repeatedly perform a task. A typical setting may be seen as consisting of a conveyor belt which transports parts, usually partially orientated and separated, a vision system obtaining information on incoming parts, and a robot performing assembly, inspection or serving other machines. Those robots are sometimes programmed on-line, guiding them through their task and storing positions and operations into a memory. The drawbacks of this method are numerous: programming errors require re-execution of programming tasks; teaching of many positions is time-consuming and error prone; programming time is not productive since all robot-related equipment has to be stopped during new task programming; synchronisation with other equipment is difficult; and sensors cannot be used to modify actions during program execution.

In recent years, there has been a significant change in the attitude of robot manufacturers, and almost every new robot is today sold with an off-line programming system. With ten years experience in the field, manipulator level languages are now accepted. Usually they are used to give the Cartesian reference frames as to where the manipulator hand should be moved in order to accomplish the task. Those frames or their sequence can be modified by run-time events, as sensor output or external synchronisation signals. In some early systems, every manipulator joint was individually given a value in its own coordinate system. However, programming in this way is not yet a good solution as it relies heavily on the detailed information given by the programmer, who in turn is expected to have some skills in mathematics and programming. Also, the actual program writing is not so easy as one might think as it is quite difficult to understand and use positions in 3D space.

An intelligent robot[1] will be given only a task and the spare parts and materials needed to perform it, instead of programs embedding the complete sequence of actions to be taken. An issue to be addressed by more advanced systems is 'world representation' and 'object modelling'. A world representation should allow the robot to manipulate parts and to sense the environment. Current industrial robots do not have any geometrical world model; their knowledge is encrypted into variables and data structures which have only meaning for their human programmer. An integration of modelling used in design and production is highly desirable. Many systems are today commercially available for CAD/CAM applications. Most of them are purely graphics systems, without much interest for robotics. The models used by CAD systems may be useful in robotics, even though robot relevant information, such as centre of gravity, mass grasping point, etc. of the object, are not contained in CAD models. Experience gained using the RAPT system is that when using CAD systems world modelling is still long and tedious. Moreover the use of CAD databases for defining world models to be used also for sensorial tasks (e.g. for vision understanding) is not yet acceptable.

Despite these formidable research issues, robots are still in use. The practical solution taken has been to limit the intelligence and the understanding of the robot, and to reduce programming of robots to usual programming or, in a few cases, to parallel programming. The use of world models is not performed in any commercial systems.

Many industrial issues in robot programming are still open. How to program a robot completely off-line, and how to express the integration of different devices, being the most studied. The present status of software and software engineering has its impact here.

Robot programming

The most obvious ways to implement robot languages are to adapt a general-purpose programming system or to develop robot-specific programming systems. In the latter case, new languages may be developed from scratch or from existing automation languages, such as APT.

The first solution is appealing because education and training of robot programmers can be reduced in time and the development of robot applications will be reduced to the writing of a few routines. The second solution is attractive because robot programmers need not be computer experts – what they have to learn will be exactly what they need to use. In the case of APT, the same NC programmers could be easily converted into robot programmers.

The first solution has as a shortcoming in that it relies heavily on what is available for general-purpose programming. The computer – user interface is not tailored on specific needs. The second solution has as a disadvantage in that many development efforts are often spent only to provide something already available with minor differences. In the case of APT-like languages, it may also be argued whether or not robot programming is the same as NC machinery programming.

Since a language is a way to express or to test different solutions, it can be said that all the developments so far obtained have demonstrated what can be done today at the manipulator level programming. After that, the choice of using existing languages or developing new ones is a problem of market image and acceptance. The need to develop new languages may not be encountered. The only important difference is in the programming environment. The development of programs off-line requires sophisticated programming environments which are not generally available in standard languages, such as FORTRAN or even PASCAL.

If the language should demonstrate new solutions or, in other words should allow task-orientated programming, then we may still want to maintain manipulator level programming as a target language for planning and sensorial activities. The expression of these activities should not necessarily resemble any of the existing programming languages. Graphics, natural language, or sometimes mathematical equations may all be useful in providing ways to express tasks.

Many papers have reviewed existing robot programming languages[2,3], and this paper is not intended to be a review of all existing robot programming systems, but to see what new ideas have emerged and whether they are assessed or waiting to be fully explored.

Usually two systems are strictly integrated in a robot programming system. The 'user language' in which application programs are written, and the 'run-time system' which executes the code generated by the language translator. Both are similar to those used, for example, in most of the PASCAL systems, it may be used as a way to standardise user languages by changing only the run-time system, as done in VAL[4], developed for PUMA robots and then implemented on other Unimation robots. Often the run-time system is run on microcomputers and written in assembly languages. This trend could change when more and more cheap computer power makes the writing of all the software in high-level languages reasonable.

An issue still to be addressed is that of defining standard software interfaces between the robot and the user level software. What kind of information is passed to the robot; joint positions or frames? In which order? How to ask for a point-to-point execution or for a continuous path?

There are no reasons why a Cartesian robot and a polar robot should be programmed in completely different ways. While the run-time system is well tailored to the specific hardware in use, the user-level language should be problem orientated rather than manipulator orientated. If this standard software interface is provided and accepted by any robot manufacturer, it may be possible to get any robot language to work for any robot; homogeneity and modularity will also be valuable in industrial settings. In the CAM-I project of standardisation in robot software[5], five main robot software components have been defined: robot language, robot simulator, robot controller, robot modeller, and teaching system. They have put AI parts in the CAD/CAM environment, and this seems more a decision not to deal now with difficult problems as the ones open in perception and decision making. In the author's opinion most of them should be solved at the robot level – the robot being the flexible and adaptable entity of the FMS.

Languages and software environments

The descriptions given in this section are based on the following key words:

- Expression of movements (joint level, hand level, object level).
- Expression of trajectories.
- Use of sensors.
- Class of the language (PASCAL, BASIC, functional).
- Implementation (interpreter, compiler, programming system).
- I/O (ports, functions provided, integration with other equipment).
- Multitasking, synchronisation, and parallelism.
- Integration with CAD/CAM for simulation, planning and control.

General-purpose languages may be looked upon as candidates for robot programming. PASCAL has already been used for robot applications[6]. However, even PASCAL presents some shortcomings as a language for automation: it does not support cooperation, which can be obtained at the operating system level, moreover, it is lousy in file management, and file management would occur very often in integrated manufacturing. Other solutions have been tried, for instance using Concurrent PASCAL, a small language developed around PASCAL to define and execute concurrent tasks.

Languages for automation did not exist before the introduction of robots. The only exception is APT, the language for NC programming. In fact, APT has been chosen as a basis for robot languages in two projects, ROBEX[7] and RAPT[8], even though its use in robotics has not yet been demonstrated as being truly useful.

Software environments for general-purpose programming are now available on most computers. The most complete and advanced software environment today are those of UNIX and INTERLISP, while the situation of ADA is not yet assessed. The main advantage of a software environment over a simple language compiler is that the former provides a unified approach for all the problems encountered in the project, the development, and the test of the program.

UNIX

The output redirection and the pipeline mechanism to connect programs are the most useful characteristics of UNIX in making modular programming easy. Different programs can be connected in any meaningful way. The UNIX operating system is very popular on the scientific personal workstations, and some of them are intended for CAD/CAM applications. UNIX could become the standard operating system for CAD/CAM applications, but perhaps not for robot programming. The computers used today to operate robots ususally run both the run-time system and the user-level language. To reduce hardware costs they are usually standalone systems, without any standard operating system. Only very sophisticated robots could justify the high cost of using a sophisticated computer.

INTERLISP

The only experience so far reported of using LISP for robot control and programming is at the MIT Artificial Intelligence Laboratory where MINI[9], an extension to LISP to deal with real-time interrupts and I/O, was used to program a robot equipped with a force sensor. LISP has not yet obtained the interest of robot manufacturers. Many reasons for this can be envisaged. Only recently LISP has been made commercially available on mini and microcomputers; it requires a lot of central memory and is inefficient in mathematical computations and array management. Despite these shortcomings, LISP can now be considered with great interest for two reasons. The functional style of programming which is the basis of LISP (even though MINI is not an example of functional programming it used a lot of SET instructions) makes it interesting because languages embedded in LISP are completely extensible, so that all the manipulation functions could be modified by the user. A second good reason for using LISP is that robot programming tends to use more and more artificial intelligence techniques, and LISP is still considered the main AI language. Serious LISP development requires several software components still not available even in the UNIX environment. This still makes LISP an intensive memory user, and is the reason for developing LISP machines to make the best use of the computer. Unfortunately those machines are still too expensive for any use at the factory level.

ADA

The main motivation for using ADA[10] in robot programming is that it is a structured and complete computer language, and offers some advanced tools such as extensibility, modularity, real-time capabilities, and strong type checking to increase the program's reliability. Moreover, it has been designed to be the only language of the 'eighties' and claim has been made that ADA could substitute any language from the assembler level to the highest levels.

The importance of a complete programming language for robotics applications has been made clear: robots should cooperate with other equipment, and this requires task synchronisation and coordination of different and sometimes non-trivial tasks. Such programming and coordination is not provided by most of the robot programming languages in industrial use today. The authors are not aware of any examples of robot run-time systems written in ADA, but there may be examples of other subsystems developed in ADA. Vision is one of them. The established practice in vision programming has been to choose C or PASCAL; ADA is a superset of them and its use should solve more problems than it can open.

Experience at the University of Michigan has demonstrated that the use of packages (a way to simplify program encapsulation) and generic packages (a way to implement abstract data types) can simplify the development of complex software, making it easier to distribute tasks to different people, to integrate them, and to modify the manufacturing cell without complex software modifications. However, the large size of ADA can be a problem. Many features of ADA are hardly useful in industrial automation, but their presence makes ADA compilers large and ADA language difficult to use.

ADA has, therefore, the ability to be a reasonable solution for programming robotic cells. The main shortcoming of this philosophy of algorithmic and explicit programming is that it is unsuitable to deal with a complicated cell in which many events may happen; time and sequence constraints can be met by different solutions. In this case expert systems, such as GARI[12] seem a more flexible and understandable way of planning and controlling the cell.

Functional languages and logic programming

Since 1980 much literature has stressed the idea that the future computer languages will be different from the present in some radical way. The evolution of languages as seen from ALGOL to PASCAL to ADA can be a deadend for computing. Those languages are based strictly on the Von Neumann architecture of computers, and that architecture is unlikely to continue in future computer generations because it has an unnecessary bottleneck in accessing memory. Languages using assignments access memory one word at a time, and assignments make user languages more suitable to the way computers operate than to the way humans think.

The next generation of computers should avoid this bottleneck. Many architectural solutions are available for that; all of them rely on using functional languages. What makes a functional language attractive is its problem-orientated expression, because it expresses functions and does not care about memory locations, it is implemented as a very restricted kernel (the function definition and composition operators) and then to grow in every way using user-defined functions, and its suitability to run on widely distributed architectures since it does not produce side effects (no global memory is used). The accent on functionality is found in many robot languages. MINI[9], LAMA-S[13], AML[14] and LENNY[15] have provided some way to obtain functional capabilities, mainly extensibility.

Another language with full functional capabilities is PROLOG. It is also the most successful language for logic programming. The Japanese Fifth Generation programs primarily relies on it as the basic language for future computers. Its use for robot prgramming has not yet been attempted. In some application fields close to robots, i.e. CAD, its usefulness has been demonstrated. In a comparison[16] between a 3D graphic program written in PASCAL and the same program written in PROLOG, the PROLOG implementation was more concise, readable and clear than the PASCAL version. It also took less storage and ran faster than the PASCAL compiled version. Since PROLOG has been used for operating systems as well as for plan generation, and has various ways for managing arrays, its applicability to robotics waits only to be fully demonstrated. It is likely that PROLOG implementation will be accepted by artificial intelligence orientated users and by mathematically orientated people.

European robot languages

The European scene of robot programming is very active. The first commercially available language for a robot, SIGLA, was developed in Europe, as are some of the most advanced projects.

HELP

HELP[17,18] is the language developed by DEA (Italy) for their PRAGMA A3000 assembly robot. It allows concurrent programming and structured programming. The syntax is PASCAL-like, and all the manipulation functions are provided as subroutines. Signal and wait provide synchronisation between different tasks. Any kind of sensor can be connected using a rich set of I/O ports operations. The robot is modular, with different arms and different degrees of freedom for each arm. The coordinate system is Cartesian, and two rotary axes can be added to every wrist. The application programs are usually provided on installation of the robot. Major applications are in the automotive industry, electronics assembly and precision mechanics. It is implemented on DEC LSI11 computers under the RT-11 operating system.

LAMA-S

LAMA-S[13] was developed by the Spartacus project – a project aimed at developing robots to help handicapped people in many every day tasks, such as serving drinks or food. LAMA-S uses APL as the implementation language. The user-level functions are translated into a low-level language, PRIMA, and then executed. Besides move instructions based on the use of frames, LAMA-S provides real-time primitives and parallel execution of tasks. The language uses two structures to define the execution order: sequence block, to indicate that all the instructions inside are to be executed sequentially, and parallel block, to indicate that all the instructions inside are initiated in parallel cobegin-coend structure). Other standard control structures are provided. The use of APL demonstrated that APL is a good implementation tool because it allows functional extensibility of the language. On the other hand it is not recommended for industrial use because of the following shortcomings: it needs an APL machine to run, the APL syntax is not convenient, the syntax analysis is not perfect, and it is difficult to implement interactive programming using APL.

LENNY

LENNY[15] is under development at the University of Genoa (Italy) for use in describing movements for an emulated anthropomorphic arm, with seven degrees of freedom. It is intended as a language powerful enough to express complex chains of actions and understandable by humans as a way to represent processes and concurrent computations. One of the key issues of LENNY is functionality. No reference can be made to any absolute kinematic quantity, they are always mechanical in context. In LENNY, the robot reference frame is fixed in the shoulder, and commands like up, down, right, etc., refer to that coordinate system. Functionality will enable LENNY to use any new procedure as part of the language.

LM

LM, Language Manipulation[19,20], was developed at the University of Grenoble (France). It is implemented on a Robitron robot (four degrees of freedom) cooperating with a Barras robot (two degrees of freedom), a Renault TH8 and a Kremlin robot (both with six degrees of freedom) and commercially available as the Scemi robot. It is PASCAL-like and frame orientated, and provides many of the features of AL except coordination and parallel execution of tasks. It is integrated with LM-Geo[21], a system used to infer body positions from geometrical relations. LM-Geo produces program declarations and instructions in LM, resembles RAPT but does not use symbolic algebraic calculus to find the frames which satisfy the equations, and analytically computes the values.

LMAC

LMAC[22] is a system for flexible manufacturing developed at the University of Besancon (France). It was designed to assure safe control of different mechanical devices in the automated cell; in doing so it performs many checks before actually executing code. It offers modularity based on the implementation of abstract data types, provides generic modules (the types of data belonging to that kind of module can be specified at run-time), and object parameterisation. External procedures written in any language can be called by LMAC programs. Different tasks representing different real-time processes can be defined and executed. Synchronisation is based on Dijkstra guarded commands. Even though its external form resembles Concurrent PASCAL it has been completely rewritten in PASCAL.

LPR

LPR, Langage de Programmation pour Robots[23], was developed by Renault and the University of Montpelier (France). It is based on defining state graphs and transition conditions. Transition conditions are used also to synchronise actions. All the graphs at the same level are executed in parallel by the supervisor; every 20ms an action from each of the same level graphs is executed. Up to 24 input/output ports can be used by LPR to provide sensor interface and synchronisation with other devices. LPR runs on a VAX

11/780 and produces code for an Intel 8086 microcomputer controlling the robot. It is available on robots produced by Renault and by ACMA Robotique.

MAL

MAL, Multipurpose Assembly Language[24,25], was developed at the Milan Polytechnic (Italy) to program a two-arm Cartesian robot evolved from the Olivetti SIGMA. It is a BASIC-like system which features synchronisation and parallel execution of tasks, as well as movement instructions and sensor interfaces. Subroutine calls with argument lists are supported. MAL is composed of two parts: translator from the input language into intermediate code and an interpreter of the intermediate code. The intermediate code is interfaced with a multimicro-hierarchical structure, and all the joints are individually driven by different microcomputers. Force sensing is also controlled by a devoted microcomputer. Photodiodes on the fingers are used as binary sensors.

PASRO

PASRO, PAScal for RObots[6] is provided by the German company Biomatik. It is based on the PASCAL language with added data types and procedures used to perform robot specific tasks. They are stored in a library and 'callable' by any standard PASCAL compiler. It is based on the AL experience. Biomatik may provide assistance in order to modify the coordinate transformations and the control interface for a new kind of robot. Procedures are provided to drive the arm point-to-point or along a continuous path. The first implementation of PASRO has been tested on a microrobot.

Portable AL

Portable AL[26] is an implementation of the AL programming environment carried out at the University of Karlsruhe (West Germany) on mini and microcomputers. It incorporates AL compiler[26], POINTY[27] and a debugging system. A dedicated operating system was developed to support I/O and multi-tasking. It runs on a PDP 11/34 and an LSI 11/2 which control the PUMA 500 robot.

RAPT

RAPT, Robot APT[8,28], in its actual implementation is an APT-like language used to describe assemblies in terms of geometric relations and to transform them into VAL programs. A RAPT program consists of a description of the parts involved, the robot, the workstation and an assembly plan. The assembly plan is a list of geometric relations expressing what geometrical relations should be held after a step in the assembly has been done. The program is completely independent of the type of the robot used. All the bodies are described as having frame position and plane, cylindrical or spherical faces. A reference system is automatically set in every feature.

RAPT builds a graph of relations and tries to reduce it to the minimum graph using a set of rules. From the reduced graph a VAL program is produced. The Computervision CADD3 system has been used to build the models and to give graphics routines.

ROBEX

ROBEX, ROBoter EXapt[7], is the off-line programming system developed at the Technical University of Aachen (West Germany) as a programming tool for FMS. Its main purposes are to develop APT for FMS and for robot off-line programming, and to be independent of the kind of robot used. Applications are in workpiece handling. APT style of programming is used to describe geometry, while the ROBEX extensions are robot movement instructions, interactions with sensors (now only binary ones), and synchronisation with peripherals (machine tools, conveyor belts, etc.). In this APT-like system for FMS three languages are used: EXAPT for NC part programming, ROBEX for part handling programs, and NCMES for measuring programs. The system is portable in two ways: it is implemented in FORTRAN IV and it generates robot independent pseudo-code which is sent to the appropriate robot for further processing and execution. The user either interactively or using a graphic interface inputs coordinates and geometry of the world and programs.

SIGLA

SIGLA, SIGma LAnguage[29,30], is the language used for programming Olivetti SIGMA robots. SIGLA is a complete software system which includes: a supervisor which interprets a job control language, a teaching module which allows teaching-by-guide features, an execution module, and editing and saving of program and data. SIGLA has been in use for many years at the Olivetti plant in Crema (Italy). Its applications span from assembly to riveting, drilling and milling. The system and the application program run in 4K of memory.

SRL

SRL, Structured Robot Language[31] (see also page 231), is under development at the University of Karlsruhe (West Germany). It is a successor of Portable AL and to some degree PASCAL. Data types as in PASCAL are added to AI data types. The declaration part also contains a specification of the system components. Instructions can be executed sequentially, in parallel, or in a cyclic or delayed way. Different motions are available, in particular straight and circular motions. The project of SRL is part of a standardisation project. The source SRL code is translated into an intermediate code, IRDATA, which is a machine independent code.

VML

VML, Virtual Machine Language[32], was developed by the Milan Polytechnic in cooperation with CNR Ladseb of Padova (Italy). Intended as an intermediate language between artificial intelligence systems and robot, it receives points in Cartesian space and transforms them into joint space. It also manages task definition and synchronisation, and is part of a hierarchical architecture, in which three levels have already been implemented.

Concluding remarks

It is hard not to get lost in the many different robot languages available. The authors have tried to review them by identifying what experience they have provided and what open problems they have not solved. Many issues have not been addressed as it was not intended to give a complete overview. Most of the new commercially available languages in the North American market have not been included. While most of the attention is now focused on acquiring manipulator level systems the authors have tried to discover what other trends and experiences are available to expand robot programming toward more ambitious tasks.

Acknowledgements

Thanks are due to J. Bach, A. Haurat, and J. Lefebre who provided unpublished material.

References

[1] Kempf, K. 1983. *Artificial Intelligence Applications in Robotics – A Tutorial.* IJCAI 83 Tutorial, Karlsruhe, West Germany.

[2] Bonner, S. and Shin, K. G. 1982. A comparative study of robot languages. *IEEE Computer*, 12: 82-96.

[3] Lozano-Perez, T. 1983. Robot programming. *Proc. IEEE*, 17(7).

[4] Shimano, B. 1979. VAL: An industrial robot programming and control system, In, *Proc. Conf. on Programming Languages and Methods for Industrial Robots.* IRIA, Paris.

[5] CAM-I proposes standards in robot software. 1982. *The Industrial Robot*, 9(4):252-253.

[6] *PASRO–Pascal for robots*. 1983. Biomatik Co., Freiburg, West Germany.

[7] Weck, M. and Eversheim, E. 1981. ROBEX–An off-line programming system for industrial robots. In, *Proc. 11th ISIR*, pp.655-662. JIRA Tokyo.

[8] Popplestone, R. J. et al. 1980. An interpreter for a language for describing assemblies. *Artificial Intelligence*, 14: 79-107.

[9] Silver, D. 1973. *The Littler Robot System*. MIT Artificial Intelligence Lab., Report AIM 273.

[10] *Reference Manual for the ADA Programming Language*. 1980. Proposed Standard Document, Dept. of Defense, USA.

[11] Volz, R. A., Mudge, T. N. and Gal, D. A. 1983. *Using ADA as a Programming Language for Robot-based Manufacturing Cells*. RSD-TR-15-83, University of Michigan, Ann Arbor, USA.

[12] Descotte, T. and Latombe, J. C. 1981. GARI: A problem-solver that plans how to machine mechanical parts. In, *Proc. 7th IJCAI*, pp. 766-772.

[13] Falek, D. and Parent, M. 1980. An evolutive language for an intelligent robot. *The Industrial Robot*, 7(3): 168-171.

[14] Taylor, R. H., Summers, P. D. and Meyer, J. M. 1982. AML: A Manufacturing Language. *Int. J. Robotics Research*, 1(3): 19-41.

[15] Verardo, A. and Zaccaria, R. 1982. *Lenny Reference Manual*. Internal Report, University of Genoa, Italy. (In Italian)

[16] Gonzalez, J. C., Williams, M. H. and Aitchison, D. E. 1984. Evaluation of the effectiveness of Prolog for a CAD application. *IEEE CG9A*, 3: 67-75.

[17] Camera, A. and Migliardi, G. F. 1981. Integrating parts inspection and functional control during automatic assembly. *Assembly Automation*, 1(2): 78-82.

[18] Donato, G. and Camera, A. 1980. A high level programming language for a new multiarm assembly robot. In, *Proc. 1st Int. Conf. on Assembly Automation*, pp. 67-76. IFS (Publications) Ltd, Bedford, UK.

[19] Latombe, J. C. and Mazer, E. 1981. LM: A high-level programming language for controlling assembly robots. In, *Proc. 11th ISIR*, pp. 683-690. JIRA, Tokyo.

[20] Mazer, E. 1983. Geometric programming of assembly robots. In, *Proc. Conf. on Advanced Software in Robotics*, Liege, Belgium.

[21] Miribel, J. F. and Mazer, E. 1982. *Manuel D'utilisation du Langage LM*. Research Report IMAG, University of Grenoble, France.

[22] Haurat, A. and Thomas, M. C. 1983. LMAC: A language generator system for the command of industrial robots. In, *Proc. 13th ISIR*, pp. 12-69. SME, Dearborn, MI, USA.

[23] Bach, J. 1983. *LPR Description*, unpublished. Renault, France.

[24] Gini, G. et al. 1979. A multi-task system for robot programming. *ACM Sigplan Notices*, 14(9).

[25] Gini, G et al. 1979. MAL: A multi-task system for mechanical assembly. In, *Proc. Conf. on Programming Methods and Languages for Industrial Robots*. IRIA. Paris.

[26] Finkel, R. et al. 1975. An overview of AL, a programming system for automation. In, *Proc. 4th IJCAI*, Tbilisi, USSR.

[27] Gini, G. and Gini. M. 1982. Interactive development of object handling programs. *Computer Languages*, 7(1).

[28] Ambler, A. P. and Popplestone, R. J. 1975. Inferring the positions of bodies from specified spatial relationships. *Artificial Intelligence*, 6: 157-174.

[29] Banzano, T. and Buronzo, A. 1979. SIGLA – Olivetti robot programming language. In, *Proc. Conf. on Programming Methods and Languages for Industrial Robots*, pp. 117-124. IRIA, Paris.

[30] Salmon, M. 1978. SIGLA – The Olivetti SIGMA robot programming language. In, *Proc. 8th ISIR*, pp. 358-363 (1978).

[31] Blume, C. and Jacob, W. 1983. Design of a Structured Robot Language (SRL). In, *Proc. Conf. on Advanced Software in Robotics*, Liege, Belgium.

[32] Gini, G. et al. 1980. Distributed robot programming. In, *Proc. 10th ISIR*, Milan, Italy.

A ROBOT PROGRAMMING SYSTEM INCORPORATING REAL-TIME AND SUPERVISORY CONTROL: VAL-II

B.E. Shimano, C.C. Geschke, C.H. Spalding III and P.G. Smith
Adept Technology Inc., USA

First presented at Robots 8, 4-7 June 1984, Detroit (SME Technical Paper MS84-448).
Reproduced by permission of the authors and the Society of Manufacturing Engineers.

VAL-II is a new robot control system and programming language. In
addition to providing the fundamental robot programming and control
features available in its predecessor, VAL, VAL-II includes: a network
communication capability which allows the system to be interfaced to a
supervisory computer; new methods for generating and modifying the
robot trajectory in real-time based upon sensory or external computer
input; computational facilities similar to those found in structured,
high-level computer programming languages; and support for concurrent
process control. Examples are presented which demonstrate how these
features can be employed to produce specialised application programs
which simplify the use of the robot.

Computers and computer-controlled machines are revolutionising manufacturing and materials handling systems. The development of low-cost,
high-performance microprocessors has made it attractive to add programmability and decision making to many automatic machines. As more
machines become computer-based, the trend has been to interface them to
communication networks to allow data transfers, centralised control, and
status monitoring.

Computer-based robots are following similar trends. When provided with
a sophisticated programming language and communication facilities, such
robots become an important part of the manufacturing system. The
communication facility allows a remote computer to control the operation of
the robot, to send data to a robot program, and to monitor the robot's
activity. It also allows the robot to receive information from various sensors.
A powerful programming language makes it easy to program complex tasks,
and allows robot motions to be described mathematically. With such a
system, it is possible to write specialised application programs which can
significantly simplify programming of a robot and its day-to-day use.

Programming languages such as AL[2], AML[4] and RAIL (see page 219) provide full computational facilities and focus extensively on applications programming.

In this paper, VAL-II[1], a new robot control system and programming language designed to enable application engineers to develop sophisticated robot control programs which require external communication capabilities, is described. Although VAL-II is largely upward compatible with its predecessor and retains the simple programming methods of VAL[3], it far surpasses VAL because of the following new features:

- Formal network communication facilities.
- Mathematical capabilities equivalent to those in a high-level computer programming language.
- An enhanced operator interface based upon a new manual control unit.
- Extended sensory interfacing capabilities.
- Facilities for real-time path modification based upon both internally and externally generated command signals.
- Facilities for performing simultaneous robot and process control activities.

In summary, VAL-II combines simple programming methods with sophisticated communications, support for complex application programming, and real-time and process control capabilities.

Design philosophy and objectives

The primary objectives in designing VAL-II were to combine formal communication capabilities, sophisticated application support, flexible robot path-control facilities, and general sensory interfaces, while making the robot system easier to program and operate.

Network communications

Robots are typically coupled to other equipment using binary signal lines to control go/no-go operations. While this method is simple to understand, it is inadequate for conveying complex information. One goal in VAL-II was to enable a remote computer system to totally supervise the operation of the robot system. This includes the ability to issue all commands normally available to an operator, to download and upload application programs, to communicate directly with application programs to provide or collect data, and to monitor the status of the system. To perform these tasks, a highly structured, formal communication protocol was developed and uniformly applied.

Our goal was to develop a network protocol which was sufficiently specialised to meet the particular needs of robotic system control and yet suitably generic to be applicable to all language-based robot systems. Also, the communication network had to be compatible with emerging technologies from the office automation and computer industries. The protocol is described briefly later.

Application support

One of the goals in the design of VAL-II was to provide a sophisticated programming language which would satisfy the needs of robot programming specialists, while at the same time providing a very easy-to-use system for those accustomed to the simplicity and convenience of teach-and-repeat robots. VAL-II allows powerful application programs to be written which can communicate with the operator via the manual control unit, a standard computer terminal, or dedicated push buttons and switches. Utilising these communication facilities, special-purpose teaching and control application programs can be written in VAL-II. When such programs are distributed with robot systems, end-users are provided with easy-to-use systems which are customised for their special needs.

To allow such programs to be written, VAL-II had to include capabilities found in standard high-level computer programming languages: real-valued variables, arrays, general arithmetic expressions, block-structured program control, and formatted input/output (I/O) to and from the system terminal and the manual control unit.

Another major goal was to greatly enhance the ability of an applications specialist to control the robot in real-time. For instance, many applications require the ability to modify the preprogrammed robot trajectory while the robot is moving. In addition, external devices must often be controlled for synchronisation and data acquisition concurrently with robot operation. To meet these requirements, one approach is to modify internally the robot control system to support each special application or sensor. This approach is quite undesirable, however, since it is time consuming, limits the rate at

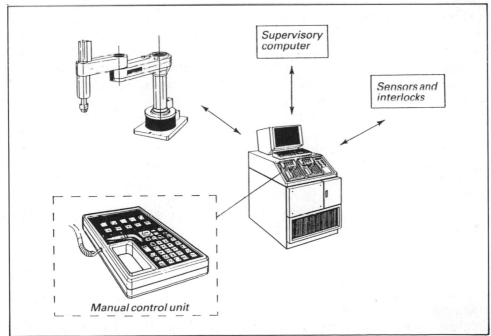

Supervisory computer

Sensors and interlocks

Manual control unit

Fig. 1 VAL-II system diagram

which new applications can be developed, and may introduce errors into the basic robot controller. Therefore, VAL-II was designed to provide general methods for sensory interfacing, real-time path control, and concurrent process control. These methods are directly accessible to the application programmer and do not require system modification.

System hardware architecture

Fig. 1 shows the major components of a typical VAL-II robot system (a four-axis, electric, direct-drive robot is illustrated). The controller cabinet houses several Motorola 68000 microcomputers, a program storage device, and all the system power and control electronics. The microcomputers are used to execute the VAL-II system-control software and application programs, to servo the motors, and to process system input and output. The controller can be used to operate up to two mechanical arms, depending upon their configuration. In addition, the controller can be interfaced to a number of different input/output devices which allow VAL-II to communicate with the outside world.

User terminal

The primary user interface to VAL-II is a standard computer terminal with keyboard – either a video terminal (CRT) or a printing terminal. The user can enter commands from the terminal to create, edit, and execute robot control programs. The user can also check on the status of the system and any programs that are executing.

Manual control

The manual control unit, or teach pendant, is a hand-held box with push buttons, lights, a numeric keypad, and an alphanumeric display. With this unit, the operator can position the robot, record positions, display and modify system parameters, and interact with a robot program.

A particularly powerful feature of the manual control unit is that it allows an operator to select, set-up parameters for, and start a robot program without using the computer terminal. Once started, the program can send messages to the operator via the alphanumeric display on the manual control and can read data entered at its keypad.

The manual control unit also allows the operator to move the robot in a variety of modes: the joints can be moved individually, the robot tool can be moved in straight lines with respect to coordinate axes fixed to either the base of the robot or the tool itself, and individual joints can be released from servo control for manual movement.

Program storage device

After robot locations are recorded and control programs have been entered, the information can be stored permanently on magnetic tape, bubble memory, or on a disk. This storage simplifies changeover from one application to another since the data and programs for the new application can be read from the storage device.

Supervisor connection

The connection to a supervisory system allows the robot system to be controlled from a remote computer, rather than from the local terminal. This capability is described later.

Binary and analogue I/O

The binary and analogue I/O lines allow application programs to monitor and control other equipment or processes in the robot workcell. This is especially important when the robot activity must be synchronised with other operations.

Binary input lines can be interfaced to simple sensors such as switches, and binary output lines can control indicators, simple actuators, and relays. Analogue lines can be used for proportional inputs from sensors, or outputs to proportional control units.

User interface

From the operator's point of view, the main part of VAL-II is the internal program which accepts commands from the keyboard or manual control unit and processes them. This program is normally referred to as the monitor since it controls all other aspect of the VAL-II system. The major monitor functions are:

- Application programs can be created or modified using a simple line orientated editor. The editor immediately checks the syntax of a newly entered line and notifies the user of any errors.
- Location variables can be interactively defined; for example, by moving the robot to desired locations with the manual control unit. Variables which are already defined can be displayed and modified.
- Once programs have been entered and associated data defined, they can be stored on the disk (or sent to the supervisory computer) and retrieved at a later time.
- An application program can be started, stopped, and restarted. A number of debugging aids are provided, as described more fully below.
- The status of an active application program can be displayed at any time.
- The value of any variable, the state of the binary I/O lines, and the current robot location can be displayed at any time.

Multi-tasking

Since VAL-II is a multi-tasking system, the user can have several different programs active at once. The monitor allows both a robot program and a process control program to be active, and still accepts commands to do things such as display status, load or save programs, or edit programs. Of course, commands which might modify an active program are not allowed.

Program debugging

VAL-II contains a number of features to facilitate program debugging. A special trace mode is available which displays each program instruction as it is executed. In addition, a single-step execution mode causes VAL-II to pause after each program instruction.

In dry-run mode, VAL-II ignores commands to move the robot. This allows programs to be tested for proper logical flow and external communication without having to worry about the robot moving to incorrect locations.

Automatic start-up

The monitor can be directed to read commands from a program instead or from the system terminal. Such 'command programs' can be used to administer standard command sequences.

If the user creates a command program which contains all the monitor commands necessary to initialise, set-up and start an application program, he can designate this program to be started automatically at power-on. This feature allows push-button control of the entire robot start-up sequence.

Supervisory communication

As manufacturing systems become more complex, there is increasing motivation for centralising the control and monitoring of factory equipment. Local area networks are being installed in factories to allow computer-based equipment to communicate for the purpose of initialisation, control, synchronisation, monitoring, and error recovery.

To this end, VAL-II provides facilities for communicating with a supervisory computer over such a network. The supervisory computer can issue monitor commands, load and store application programs and data, and monitor the status of the VAL-II system. With appropriate software, the supervisory system can completely replace the local terminal and local disk storage device. Thus, a single supervisor, with a large mass storage device, can eliminate the need for local terminals and disk drives at a number of robot systems.

Network interface

The network software is implemented using a three-layer approach, with each layer handling a different aspect of the communication protocol:

- The bottom layer performs physical I/O and handles any communication errors.
- The middle layer acts as a multiplexor to allow messages for several I/O processes to be sent over a single physical line.
- The top layer identifies the various I/O requests to the supervisory computer.

The specific I/O processes which are supported are described below.

Network I/O processes

During non-network operation, the robot system terminal is used for monitor commands and terminal I/O for application programs, and the local disk is accessed for disk operations. When the supervisory communication interface is active, any combination of these three types of communication can be redirected over the network. The I/O operations which are not redirected continue to operate normally.

Monitor I/O. When monitor I/O is redirected to the network, VAL-II accepts monitor commands from the supervisor. Output generated by commands, and error messages are returned to the supervisor. This mode allows the supervisor to completely control the operation of the robot system.

Program I/O. Program I/O requests can be redirected to the network, independent of monitor I/O redirection. Once redirected, program I/O instructions access the supervisor instead of the local terminal.

Disk I/O. When disk I/O is redirected to the network, all requests to load and save programs and data are sent to the supervisor. This allows programs which have been prepared off-line to be downloaded to robots for execution.

Status monitoring. In addition to the above redirectable processes, a special monitoring process is available only over the network. At any time, the supervisory computer can request VAL-II to report the status of the robot system. Normally, the supervisor requests this status information every few seconds.

VAL-II language: Computation and control

VAL-II supports computational facilities common to a number of high-level programming languages. Of course, it also includes special features for robot control. Since VAL-II is similar to VAL, a programmer familiar with VAL can immediately write programs, using a subset of the entire language. Major features of the VAL-II language are:

- The data types are real-valued scalars, transformations (which describe locations in Cartesian coordinates), and joint locations (which record the actual position of each joint).
- Normally, program variables are global to all programs loaded in memory. However, variables can be declared to be local, meaning that they are known only within the program in which they are used.
- One-dimensional arrays of any data type can be used. The size of an array is not explicitly specified, and the array grows automatically as its elements are referenced. This feature is particularly useful when a robot path is defined by an arbitrary number of locations.
- General arithmetic and logical expressions can usually be used wherever a real or logical value is required. These expressions can include arithmetic, logical, and relational operators.
- Standard system functions, as shown in Tables 1 and 2, provide computational aids as well as robot status and internal VAL-II information.

Table 1   Partial list of real-valued functions

Arithmetic functions

ABS	absolute value
BCD	binary coded decimal equivalent
DCB	decimal equivalent of a BCD value
FRACT	fractional part of a real number
INT	integer part of a real number
SQR	squared value
SQRT	square root

Trigonometric functions

ATAN2	arc-tangent given two numbers
COS	cosine
SIN	sine

Location functions

DISTANCE	distance between two locations
DX	x component of a transformation
DY	y component of a transformation
DZ	z component of a transformation
INRANGE	indicates if location within reach

Status information functions

ERROR	code number of last error message
RLAST	number of last element in an array
SPEED	motion speed setting
STATE	current execution state

Hardware functions

ADC	reading from A-to-D converter
HAND	current hand opening
ID	robot model, serial number, etc.
PENDANT	teach pendant button states
SIG	state of binary I/O bit(s)
TIMER	timer value

Table 2   Partial list of location-valued functions

Informational functions

BASE	robot base-to-world relationship
DEST	planned destination of last motion instruction
HERE	current tool-tip location

Computational functions

| SCALE | transformation with rescaled position vector |
| SHIFT | transformation with offset position vector |

Location variable constructing functions

| FRAME | transformation defined by four transformations which specify a reference frame and orientation |
| TRANS | transformation defined by six numeric values |

- Program control structures similar to PASCAL are supported, including:

```
IF ... THEN ... ELSE ... END
WHILE ... DO ... END
DO ... UNTIL...
FOR ... END

CASE ... OF ... END ...
IF ... GOTO ...
GOTO ...
```

- Programs can be called as subroutines and can be passed arguments of any data type.
- Text and formatted numeric output can be sent to the user terminal or the manual control unit. Numeric input can be read from either device.

In addition to these basic features, VAL-II contains a number of special features used for controlling the robot and interfacing with external equipment.

Compound transformations

It is often desirable to define robot locations relative to a reference frame other than that of the robot. For example, when parts are transported on pallets, the locations of the parts relative to the pallets can be considered as constants even if the pallets themselves have varying locations. Once the location of a pallet is known (e.g. from sensor information), a 'compound transformation' can be used to define the location of a part on the pallet in the robot reference frame.

For example, if *pallet* is a transformation which defines the location of a pallet relative to the robot coordinate system, and *part 1* is a transformation which defines the location of a part relative to the pallet, then the compound transformation *pallet : part 1* defines the location of the part in the robot reference frame.

Compound transformations can consist of any number of 'relative' transformations linked together in this manner. The transformation value resulting from evaluation of a compound is determined at the time the compound is encountered during program execution. Thus, the elements of compound transformations can have constant values, or they can vary from one evaluation to the next.

Belt variables and conveyor tracking

A part which lies on a conveyor belt can also be thought of in terms of a compound transformation. The actual location of the part can be determined if the location of the part relative to the belt and how the belt is moving is known. If the location of the part relative to the belt is sensed (e.g. with vision, photocells or switches) and the belt is instrumented to detect its motion, then the robot system can determine the part location and track it.

VAL-II programs can specify belt tracking by means of special 'belt variable' transformations. The location defined by such a variable is linked

```
        IF SIG(no. part) THEN          ; If part not ready,
            SIGNAL feed.part           ; feed a part and
            WAIT SIG (part. ready)     ; wait for it
        END
        MOVE get.part                  ; Move to the part
        CLOSEI 0.0                     ; Grasp it
```

Fig. 2   Process synchronisation

to the motion of a conveyor belt and changes as the belt moves. Thus, when the robot is instructed to move to a location relative to a belt variable, it automatically tracks the belt as long as the location is within reach.

Process synchronisation

Robot applications frequently require coordination of robot motions and actions with the cooperation of other equipment in the robot workcell. VAL-II provides a group of program instructions which can be used to achieve such coordination.

As an example, Fig. 2 shows an instruction sequence which synchronises the robot with a parts feeder. If an external signal indicates that there is no part to be picked up, the program signals the parts feeder to dispense a part and then waits for it to be in place before robot motion resumes.

To allow the use of timing information in control relationships, VAL-II provides a number of timers which can be set and read at any point in a program.

In addition to interactions with other equipment in the robot workcell, the synchronisation operations can be used to coordinate the robot control program and the process control program. This ability adds to the versatility of the system for general process control.

Exception handling

VAL II permits the programmer to associate a subroutine with an external signal or error condition so that the subroutine is called whenever the condition occurs. Such a subroutine call is referred to as a program exception since it interrupts the normal program flow.

This feature is particularly powerful, since it allows a program to monitor and react to asynchronously occurring conditions, without having to test for them explicity. Upon returning from the exception subroutine, the application program can continue operation at the point where it was interrupted. Program exceptions can be specified to monitor a variety of conditions, such as:

- State changes of binary input signals (either positive or negative transitions).
- Moving-line window violations; that is, while being tracked by the robot, a part on a moving conveyor belt moves out of a program-defined working range.
- VAL-II program errors or hardware failures. (An exception subroutine can attempt to recover from an error or perform a controlled shutdown.)

Robot motion control

In almost all robot control systems, the motion of the robot is specified as a series of locations to which the robot is to move. Even in applications where the path of the robot tool tip is of primary concern, such as in spray painting, the path is specified by many closely spaced locations. In other applications, such as spot welding, only a few locations are specified and the path between them is established by the control system.

VAL-II uses an interpolation function to automatically generate a series of intermediate locations between specified initial and final locations, thus moving the joints in a coordinated, predictable fashion between taught locations. Two different interpolation schemes are used: joint and Cartesian straight-line.

Joint-interpolated motions are generated by interpolating the joint positions from their initial values to their desired final values so that all the joints complete their motions simultaneously. Cartesian-interpolated motions are generated by interpolating the Cartesian tool-tip location and computing the joint positions necessary to move the robot tool tip along a straight line.

In addition, VAL-II automatically generates transitions between successive motion segments to produce a smooth, continuous motion defined by multiple locations. In this way, any combination of joint and Cartesian interpolated motions can be blended together to produce an arbitrarily long continuous motion. The ability to generate continuous path motions of this type reduces cycle time and is essential in certain processes such as arc welding and sealant application.

Motion instructions

A set of program instructions command the robot to make various joint-interpolated or straight-line motions. These instructions generally specify a new location to which the tool tip of the robot is to be moved. The location of the tool is a combination of the location of the robot base, the location of the end of the robot with respect to its base, and the location of the tool tip with respect to the end of the robot. The various components of the tool-tip location can be specified independently to allow for changes in the robot base location or use of different tools.

The programmer can also specify various parameters used in the robot path control algorithm. These include:

- The speed at which motions are performed.
- The final error tolerance which must be achieved before a motion is considered complete.
- The method by which steady-state position errors are corrected.
- Whether or not continuous-path smoothing should be performed between motion segments.

Real-time external trajectory modification

VAL-II programs normally assume that the programmed locations and the paths to be followed are completely predictable. For applications which involve some degree of uncertainty, sensors can be added which provide the robot with additional information to modify its path while the motion is in progress.

The use of sensory input to dynamically modify the robot motion is called 'real-time path modification'. In VAL-II, it is also referred to as 'alter mode', since nominal programmed robot locations are altered by the sensory input. While in alter mode, all robot motions are automatically modified by the input data.

Alter modes. Path modification data is passed to VAL-II as Cartesian coordinate components and can only be applied to straight-line robot motions. The data can specify any combination of offsets along and rotations about the x, y, and z axes.

Upon entering alter mode, the programmer specifies the coordinate system for path modification as either the world or tool coordinates (fixed to the robot base and tool, respectively). All received data then causes tool-tip motions to be modified relative to that coordinate system.

The programmer also selects whether or not the effects of the path modification data are to be cumulative. In non-cumulative mode, only the most recent input data affects the tool-tip location. In cumulative mode, the effects of each data message are summed and retained, so that the tool-tip location reflects all past data.

Sensory input interface. Path modification data is obtained from one of two sources: it can come from an external computer, or it can be computed by a VAL-II process control program.

While external alter mode is active, the system sends messages to the external computer once during each path interpolation cycle. The external computer must respond by returning data describing how the nominal robot tool trajectory is to be modified. The messages from VAL-II also inform the external computer of the current tool-tip location and robot status. Communications with the external computer occur over a serial line, using a simple error-detecting protocol.

Procedural motions

As described earlier, the ability to move along straight lines and joint-interpolated arcs is built into the basic operation of VAL-II. Sometimes it is desirable to move the robot tool along a path which is algorithmically or mathematically described. VAL-II allows such motions to be performed by permitting a program to compute the robot trajectory as the robot is moving. Such a program is said to perform a 'procedural motion'.

Actually, a procedural motion is nothing more than a program loop which computes many short motions and issues the appropriate motion requests. Parallel execution of robot motions with non-motion instructions allows the motion to be computed without stopping the robot. The transitions between computed motion segments are automatically smoothed by the continuous-path feature.

The following is an example of a procedural motion. This program segment moves the robot tool at a constant speed along a circular arc. The path is not prerecorded; it is described by the radius and centre of the circle to be followed. It is assumed that the real variables *start*, *last* and *angle.step* have been defined to describe the portion of the circle to be traced, that a real variable *radius* has already been assigned the radius of the desired arc, and that *x.centre* and *y.centre* have been assigned the respective coordinates of the centre of curvature. Finally, a transformation *cframe* is assumed to have been defined to describe the plane of the circle.

```
FOR angle = start TO last+start STEP angle.step
       x = radius*COS(angle) + x.centre
       y = radius*SIN(angle) + y.centre
       MOVE cframe:TRANS(x, y, 0, 90, -90, 0)
END
```

When this program segment is executed, the *x* and *y* coordinates of points on the circle are repeatedly computed. They are then used to create a transformation which is applied relative to the reference frame *cframe*.

Detached motion control

Traditionally, robot programming systems have provided a single method for teaching robot locations. The operator uses a manual control unit to move the robot through a sequence which is later replayed under automatic control. At each critical location, control parameters are selected and a record button is pressed to record the information. Such fixed methods are simple to use and sufficient for the tasks performed by simple teach-and-repeat systems. However, for more complicated tasks, involving multiple options and sensory feedback, more sophisticated teaching aids are required.

To facilitate the development of specialised teaching aids, VAL II allows an application program to release control of the robot. While the robot is detached, all the standard manual control modes are available to the operator moving the robot. At the same time, the program can continue to run, prompting the operator, and monitoring the manual control unit. Thus, complicated calibration or teaching sequences can be reduced to a series of simple steps directed by a program. At each step, the effect of selected buttons on the manual control can be redefined to maintain a simple push-button operator interface.

Concurrent process control

In typical robot installations, there are often several machines or processes which must be coordinated. Various binary and analogue status and control lines must be connected, along with switches and lights for an operator interface. The solution is usually to add a programmable logic controller (PLC) to the system, and interface it to the various machines and to the operator control panel. To avoid the need for a PLC, and to facilitate interfacing to the robot, VAL-II allows a second application program to run concurrently with the main robot control program for the purpose of process control.

```
; MOVE PARTS FROM A PART FEEDER TO PALLETS ON AN INDEXING
CONVEYOR

ENABLE NETWORK                       ; Connect to the supervisory interface

RESET                                ; Reset all the output signals

: Assign names to external signals

new.pallets = 1                      ; Output signal to the conveyor
more.pallets = 1001                  ; Input signal from the conveyor
pallet.set = 1002                    ; Input signal from the conveyor

; Ask the operator for the type of pallet to be processed

PROMPT "Enter code for type of pallet to be filled:", code

count = 0                            ; Initialise the pallet counter

TIMER 1 = 0                          ; Reset the process timer

WHILE SIG (more.pallets)             ; Process pallets until all are done

    SIGNAL new.pallet                ; Feed the next pallet
    SIGNAL -new.pallet               ; Turn off the pallet advance
    WAIT SIG(Pallet.set)             ; Wait for the pallet to be in place

; For each pallet position : compute location and move a part to it

FOR row = 0 TO row.count[code]-1     ; Loop through all the rows
    FOR col = 0 TO col.count [code]-1 ;    and all the columns

    SET put = SHIFT(frame[code] BY row*r.space[code], col*c.space[code], 0)

CALL move.part (get, put)            ; Follow a subroutine to move the part

    END                              ; END of the "col" FOR loop
END                                  ; END of the "row" FOR loop

count = count + 1                    ; Count the completed pallet

END                                  ; End of the WHILE loop

; All pallets done--output performance report to supervisory system

ENABLE REMOTE.PIN                    ; Direct output to the supervisor

TYPE /C1, "Robot", ID(1),"-", ID(2), "has completed",/S
TYPE count, "pallets (", count*row*col, "parts loaded)"
TYPE "Elapsed time was", TIMER(1)/60, "minutes"

DISABLE REMOTE.PIN                   ; Direct output back to the terminal
DISABLE NETWORK                      ; Disconnect from the supervisor

STOP                                 ; All done
```

Fig. 3   Example palletising program

A process control (PC) program can execute all the standard VAL-II program instructions, except for instructions which directly cause robot motion or action. It can read and write binary and analogue I/O signals, read and modify program variables which are shared by the main robot control program, and use all the powerful computational facilities of VAL-II. When active, a PC program runs concurrently with the main robot program and is guaranteed a certain percentage of each computer cycle, so that the worst-case response time of the PC program is predictable.

The process control program can interact with the robot program in limited but powerful ways. First, it can stop the current robot motion immediately. Second, it can dynamically modify the nominal robot trajectory in a manner analogous to the real-time external trajectory modification feature. That is, a process control program can be written to compute the amount of trajectory modification required, and send it to the trajectory generation routines every major interpolation cycle. Such a program could, for example, read an analogue input and use the data to modify the nominal tool-tip path, while the main robot program moves the tool along a complex trajectory consisting of many predefined motion segments.

Synchronisation between the main robot control program and the process control program can be accomplished by using shared program variables as flags, or by using internal binary signals.

```
; SUBROUTINE move.part -- PICK UP A PART AT pick AND PUT IT
DOWN AT place
;
; Calling sequence: CALL move.part (pick, place)

ARG LOCATION pick, place           ; Declare passed location
                                     parameters
LOCAL LOCATION pick, place         ;   to be local variables

; Get a part from the source location

APPRO pick, 30                     ; Approach the part
MOVES pick                         ; Move to the part
CLOSEI 0                           ; Grasp the part
DEPARTS 50                         ; Back off

; Deposit the part at the destination location

APPRO place, 50                    ; Approach the final location
MOVES place                        ; Move to the location
OPENI 25                           ; Release the part
DEPARTS 30                         ; Back off

RETURN                             ; Return to calling program
```

Fig. 4   Example motion subroutine

Example program

A sample VAL-II application program, presented in Fig. 3 is now considered. The application task is to load parts from a feeder into pallets on an indexing conveyor. The part locations in the pallets form a pattern of evenly spaced rows and columns. (Although the program should be more robust for a real application, it illustrates the use of several VAL-II features.)

The program begins by activating the communication link to the supervisory system. Next, all the binary output signals are turned off and names are associated with the binary I/O signals used by the program.

The program requests the operator to enter a code number at the terminal to identify the type of pallet to be loaded. This provides a means for the same program to be used for different production runs requiring different types of pallets. The value entered by the operator is stored in the variable *code*, which is used to select the appropriate elements from arrays of data describing the pallets. The arrays describe the number of rows and columns of part positions (*row.count* and *col.count*), the row and column spacings (*r.space* and *c.space*), and the location in the robot reference frame of the first part position on the pallet (*frame*). (These arrays must have been entered into the system previously.)

A WHILE...END structure forms the main processing cycle. It directs the program to continuously process pallets until an external input signal (*more.pallets*) is turned off. For each pallet cycle, the program uses binary signals to activate the indexing conveyor (*new.pallet*) and to tell when a new pallet is in position (*pallet.set*).

Nested FOR...END loops are used to index through the rows and columns of the pallet. For each part, the next location in the pallet is calculated by shifting the reference location on the pallet by the appropriate number of row and column spacings. The subroutine *move.part* (shown in Fig. 4) is called to direct the motion of the robot between the part feeder and the calculated location.

The subroutine receives the source and destination locations as arguments passed from the main program, and treats the data as local variables. Thus, this same subroutine could be used by any other program which needs to have the robot move an object from one location to another.

After all the pallets have been loaded, the program temporarily redirects terminal output to the supervisory network. This permits the program to send a performance report consisting of the robot identity, the number of pallets and parts processed, and the total processing time. The message has the following format:

```
Robot xxx-xxx has completed xx pallets (xxxx parts loaded)
Elapsed time was xx minutes
```

Acknowledgements

The authors gratefully acknowledge the contributions made by the following individuals: Prof. Richard Paul, who developed many of the fundamental concepts contained within VAL and VAL-II; Brian Carlisle, who contributed many ideas and helped in establishing the design goals; Edison Hudson and Richard Casler, who made numerous suggestions; and Joseph Engelberger, who supported the initial development effort.

References

[1] *User's Guide to VAL-II*. Adept Technology Inc., Mountain View, CA, USA, 1984.
[2] Mujtaba, S. and Goldman, R. 1979. *AL User's Manual*, Memo AIM-323. Stanford Artificial Intelligence Laboratory, CA, USA.
[3] Shimano, B. 1979. VAL: A versatile robot programming and control system. In, *Proc. COMPSAC 79*, Chicago, pp. 878-883.
[4] Taylor, R., Summers, P. and Meyer, J. 1982. AML: A Manufacturing Language. *Int. J. Robotics Research*, 1(3): 19-41.

PROGRAMMING VISION AND ROBOTICS SYSTEMS WITH RAIL

J. W. Franklin and G. J. VanderBrug
Automatix Inc., USA

First presented at the Robots 6 Conference, 2 – 4 March 1982, Detroit (SME Technical Paper MS82-216). Reproduced by permission of the authors and the Society of Manufacturing Engineers.

Many robots and industrial vision systems are programmed by the teach-by-showing method. This method is easy to use and is adequate for many applications. However, other applications are easier to solve by off-line programming. Off-line programming is more flexible and allows interfacing to other databases, such as computer aided design systems. It integrates robots and vision systems into the total manufacturing process. The RAIL language and the A132 controller, provide capabilities to program robots and vision systems using both the teach-by-showing and the off-line programming methods.

The most widely used method for programming industrial robots is the teach-by-showing method. Here programming is accomplished by manually positioning the robot at each of the locations required to perform the task. Values for each of the locations are stored in a database. Usually these values are stored in terms of the robot's individual joints, although some systems store them using a Cartesian representation.

Often the robot is moved from part to part with a manual control box, with a RECORD button to indicate which positions are to be stored in the database. Some applications areas, such as paint spraying, require long paths and smooth motions. For these areas the robot is usually physically moved through the desired motions while the control system builds up the database by sampling the positions of the joints. Usually, values which indicate various actions of the robot are associated with individual points. These values can include: time delays, sending/receiving signals, tool manipulation such as opening/closing a gripper or turning on/turning off a welding arc, and whether the robot should come to a stop at the point or move through the points with smooth motion. These values are usually associated with points by setting switches during the teach sequence. During execution, the robot plays back the point values along with the associated actions.

The teach-by-showing method is sufficient for many applications, but for some it is less than ideal. Some applications, including many in the automotive and aerospace industries, have hundreds of points. It is tedious and error prone to teach these points individually. In addition, it is a waste of resources to teach the points manually, since often this information exists in a CAD/CAM database. Off-line programming, which allows using a CAD/CAM database, offers many advantages. It provides the opportunity for the robot to become more integrated into the total manufacturing system. It also allows flexibility in scheduling for various workstations.

Another benefit of off-line programming is the ability to set offset values, and hence eliminate the necessity of reteaching when fixtures or parts in the work environment have been (slightly) moved. In addition, off-line programming makes it possible to integrate real-time sensory data into the operation of the robot, and thus allow it to work in less structured environments.

Programming vision systems

Somewhat analogously to industrial robots, most industrial vision systems are also programmed with a teach-by-showing method. Normally a vision system is capable of computing a large set of features of an image and has been programmed to perform pattern recognition on these features. Techniques frequently used include nearest neighbour and binary decision tree classification. In an inspection application, for example, the user chooses a subset of the features and, during teach mode, shows the system a

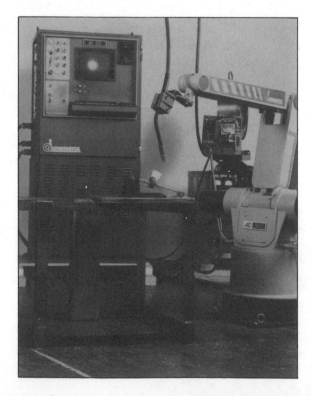

Fig. 1   The AI32 controller along with the AID 800 robot

set of good and a set of bad parts so that a statistical database can be built up. During execution this database is used to classify an unknown part as either good or bad.

In off-line programming for vision systems a language is used to write a program to perform the inspection task. In doing so, this program explicity uses part data, such as geometry. As for robot programming, this data frequently exists in CAD/CAM databases, and off-line programming makes it possible to use it for vision tasks.

Other advantages of off-line programming for vision are also similar to the advantages of off-line programming for robots. It allows for greater integration into the manufacturing systems, provides for flexibility in workstation scheduling, and allows for more efficient use of resources.

RAIL is a language which has been developed to provide a flexible interface to vision and robotic systems. It provides capabilities to program these systems using the teach-by-showing and the off-line programming methods.

The language is powerful yet easy to use. It is executed interpretively and provides useful tools for program debugging. It has been used with visual inspection systems, robotic arc welding systems, and robotic assembly systems.

AI32 controller

RAIL application programs run on the AI32 controller (Fig. 1). It includes a single-board computer, the CP32, based on the Motorola 6800 microprocessor chip and provides for 32-bit arithmetic operations and 24-bit addressing. It can directly address up to 16 million bytes of memory. A specialised enclosure protects the AI32, its power supplies, and other optional items from adverse factory conditions such as dirt, dust, electrical noise and temperature extremes.

The Autovision II, shown in Fig. 2, is a visual inspection system based on the AI32 architecture. It includes a fast, bit-slice preprocessor board to interface to cameras, store picture frames, and perform low-level picture operations. This vision processor runs in parallel, so that vision processing can be overlapped with execution of RAIL application programs on the CP32.

The RAIL language, robot control software, and image processing software run under an Automatix developed operating system, which includes an editor, a file system on cartridge tapes, communication interfaces to serial and parallel ports, and a multiprocessing environment. It provides allocating or releasing memory, starting or stopping processes, sending or receiving messages, and awaiting or causing events. These capabilities form a basis for the implementation of the system software.

User interfaces

The AI32 provides four ways for an application engineer or operator to use the system: a control panel, an interactive command module (ICM), a keyboard/console, and an RS232 port for communicating with other computers.

Fig. 2   The Autovision II
inspection unit

The control panel is the simplest of the four and can be used to power up the controller, run existing application programs, and activate some safety features. Most applications can be set-up so that the day-to-day operation of the system requires only the use of the control panel.

The ICM is a hand-held device used for moving the robot and creating, modifying, and running application programs. It has an alphanumeric display that allows the user to specify instructions and define parameters in a menu-driven manner. As the user presses buttons on the ICM, the system will automatically generate the RAIL language statements required to implement the program.

The operator console is used to enter RAIL commands and gives access to the full power of the controller. The console provides some standard tools for writing programs such as an editor, a file system, and debugging aids.

Finally, the RS232 line can be used to connect the AI32 to another computer system. This line can be used to transmit CAD/CAM location data, production statistics, and parts inventory data.

RAIL language features

RAIL is a procedural language for vision and robotic applications. The language provides a powerful set of tools for these applications, including:

- Commands for moving the robot, approaching or departing from locations, and operating a gripper.

- Robot welding commands and facilities for setting welding parameters such as voltage and wire feed rate.
- Image processing commands for taking pictures, analysing objects for such things as number of holes, object area and orientation.
- Data types include integers, real numbers, character strings, logical data, arrays, frames, and relative transformations.
- Variable names up to 20 characters long may be used for such things as locations, weld paths, program counters, or production statistics.
- Program control structures similar to those provided by the Pascal language, including IF THEN ELSE, REPEAT UNTIL, DO WHILE, WAIT, and functions. These structures permit an application program to make decisions at runtime, exchange the sequence of execution of instructions, or repeat the execution of any instructions.
- Arithmetic, comparison, and logical expressions. Built-in functions for square root, sine, cosine, arc-tangent, absolute value, time-of-day, and interval timer.
- A simple way to access input or output lines connected to other factory equipment such as part fixtures, part detector switches, conveyor belts and welding positioners.

The RAIL user is provided with a workspace for storing user programs and variables. Programs can be created, displayed, edited or deleted. Commands are provided for storing variables and programs on the file system and for retrieving them later.

RAIL is implemented as an interpreter. When source statements in RAIL are typed in at the console or loaded from tape, they are translated into an intermediate code. The intermediate code permits RAIL programs to execute efficiency, while still providing the benefits of an interactive debugging environment. The translation of RAIL statements into intermediate code is done at high speed, so user statements appear to be executed as soon as they are typed in.

Interactive command module

The interactive command module (ICM) is a hand-held device through which the user can communicate with the robot and AI32 controller. It weighs about 0.5kg and can easily be held in one hand (see Fig. 3). The ICM capabilities include:

- Moving the robot in joint, world, and tool motion.
- Defining location data such as points, paths, and frames.
- Creating RAIL programs with a subset of the RAIL language.
- Editing the RAIL programs created from the ICM.
- Controlling the execution of RAIL program, including start, stop at the end of a cycle, suspend, or single step.
- Defining and altering robot and application parameters, such as robot speed and welding schedules.

As an example of the use of the ICM, suppose the user wants to make the application program wait until an input line is turned on. When the WAIT INPUT button on the ICM is pressed, RAIL will use the ICM display to

Fig. 3   The interactive command module (ICM), a hand held device for communicating with the AI32 controller and the robot

prompt for the input line to be tested, and the desired state of the line (on or off). It will then generate a WAIT UNTIL statement that will test the specified input line and insert it into the application program.

For many robot applications, where the level of control and the amount of logic in the solution is not too great, the user can develop the entire program from the ICM. This eliminates typing at the console and greatly simplifies the programming task.

Sample RAIL programs

Figs. 4–7 show sample RAIL programs for inspection, welding and assembly. The programs illustrate the general features of the language, such as vision system commands, robot motion, and program control structures.

The RAIL program shown in Fig. 4 is an inspection program for mounting brackets moving along a conveyor belt. There are input signals from the conveyor and a part detector. BRACKET waits for the next part, takes a picture, and turns output signal GOOD_PART (the name of the output signal) on or off depending on whether the area of the part is within 20 of the nominal area BR_AREA.

For simplicities sake this program uses only part area as a criterion for accepting or rejecting parts. RAIL supports many other inspection criteria, including number of parts, number of holes, part size and orientation, part brightness, and comparison with known good or bad parts.

```
FUNCTION BRACKET
  BEGIN
;
; BRACKET inspects mounting brackets
; moving along a conveyor belt.
;
    BR_AREA = 1623
    WHILE CONVEYOR == ON DO
      BEGIN
        WAIT UNTIL PART_READY == ON
        PICTURE
        IF OBJ_AREA WITHIN 20 OF BR_AREA THEN
          GOOD_PART = ON
        ELSE
          GOOD_PART = OFF
      END
  END

INPUT PORT CONVEYOR 1
INPUT PORT PART_READY 2
OUTPUT PORT GOOD_PART 1
```

Fig. 4   A RAIL inspection program for mounting brackets moving along a conveyor

Fig. 5 shows two RAIL programs for a welding application. The program WELD_PARTS waits for the next part, clamps it down, welds it, and then unclamps it. After ten parts it will call the second function CLEAN_ TORCH to clean the torch nozzle.

The RAIL statements in Fig. 6 illustrate the use of relative transformations to perform a three-pass weld. Each pass is 2.0mm above the previous weld. The METRIC switch is used to tell RAIL that the offsets and the approach/depart distances are in millimetres.

An assembly program for inserting board spacers into a pcb is shown in Fig. 7. The program takes spacers from a feeder, applies glue to the bottom, and then inserts the spacers into the spacer holes in the board. The position of the spacer holes is relative to the position of the board. If the board is moved then only the board position PCBOARD must be retaught, not the holes.

Interfaces to other databases

For robots to become an integral part of a total manufacturing system it is essential that they be interfaced to other databases. Such interfacing removes the time-consuming and error-prone job of teaching, allows more efficient use of resources, and provides for flexibility in scheduling of workloads to various robotic workstations. In addition it helps to reduce the time lag between design of a product and its availability in the market place.

```
FUNCTION WELD_PARTS
  BEGIN
    FOR PART_NO = 1 TO 10 DO
      BEGIN
;
; Move to HOME position, wait for next part,
; clamp it down, weld it, and unclamp.
;
        MOVE SLEW HOME
        WAIT UNTIL PART_READY == ON
        CLAMP = ON
        APPROACH 2.0 FROM PART
        WELD PART WITH SPEEDSCHED[2], WELDSCHED[5]
        DEPART 2.0
        CLAMP = OFF
      END
;
    CLEAN_TORCH
  END


FUNCTION CLEAN_TORCH
  BEGIN
;
; Brush out the torch nozzle, then spray it.
;
    APPROACH 2.0 FROM CLEANER_BRUSH
    BRUSH = ON
    MOVE CLEANER_BRUSH
    DEPART 2.0
    BRUSH = OFF
;
    MOVE CLEANER_SPRAY
    SPRAY = ON
    WAIT 2 SEC
    SPRAY = OFF
    DEPART 2.0
  END

INPUT PORT PART_READY 1
OUTPUT PORT CLAMP 1
OUTPUT PORT BRUSH 2
OUTPUT PORT SPRAY 3
```

Fig. 5   Two RAIL programs for a welding application. The first program welds the parts, the second cleans the torch

```
;
; Perform a three pass weld, each pass 2.0 mm
; above the previous weld.  The variable BACKUP2MM
; is used to calculate the offset from the weld seam.
; On each pass the robot approaches to 50 mm above
; the start of the weld, performs the weld using the
; current offset, and then departs 50 mm from
; the weld.
;
    METRIC = ON
    OFFSET = 0
;
    FOR PASS = 1 TO 3 DO
      BEGIN
        BACKUP2MM = [ 0, 0, -OFFSET, 0, 0, 0 ]
        APPROACH 50.0 FROM SEAM
        WELD SEAM:BACKUP2MM WITH
                     SPEEDSCHED[3], WELDSCHED[1]
        DEPART 50.0
        OFFSET = OFFSET + 2
      END
```

Fig. 6   A RAIL program for performing a three-pass weld

The most important database for interfacing to a robot is the design database in a CAD/CAM system. These databases contain high amounts of information that (in effect) are regenerated in the teach-by-showing method of programming. Examples include an airplane wing skin that has hundreds of holes of various sizes and locations, automotive spot welding with many car models each with many welds, and subcomponent arc welding with numbers of different parts each with many welds. Each of these examples illustrates the potential productivity gains of robotic interfaces to CAD/CAM systems and demonstrates the importance of working toward making these interfaces a reality on the shop floor.

In some applications the program logic (i e. the sequence of tasks and the types of error recovery) will reside in the robotic system, and the nature of the data transferred from the CAD/CAM system will be primarily geometric. For example, a robotic visual inspection system may store the inspection procedures (including the particular image processing used in the inspection programs), with the CAD/CAM system transferring to the robotic system the geometric data (such as edge locations, hole locations, hole sizes, tolerances) of the parts to be inspected. In other applications an entire robotic program may be transferred from the CAD/CAM system. For example, in an arc welding application the design/manufacturing engineer may specify the welding process information to the CAD/CAM system so that an entire robotic program can be generated. In still other applications there are other databases that will play a role. For example, a process planning database has information on the nature and sequence of tasks in assembly operations. A process planning database will clearly play an important role in integrating robotic assembly stations into a total manufacturing system.

```
    FOR SPACER = 1 TO LAST_SPACER DO
        BEGIN
;
; Get next spacer from feeder.
;
        OPEN
        APPROACH 2. 0 FROM FEEDER
        MOVE FEEDER WITH SPEEDSCHED[3]
        CLOSE
        DEPART 2. 0
;
; Put glue on end of spacer.
;
        APPROACH 2. 0 FROM GLUE
        MOVE GLUE WITH SPEEDSCHED[3]
        SQUIRT = ON
        WAIT 0. 5 SEC
        SQUIRT = OFF
        DEPART 2. 0
;
; Insert spacer into next spacer hole in PC board.
; SPACER_HOLE[SPACER] is the offset of the hole
; from the corner of the PC board.
;
        HOLE = PCBOARD: SPACER_HOLE[SPACER]
        APPROACH 2. 0 FROM HOLE
        MOVE HOLE WITH SPEEDSCHED[3]
        WAIT 1. 0 SEC
        OPEN
        DEPART 2. 0
    END
```

Fig. 7   A RAIL program for inserting board spacers in a pcb

For robotic systems to become fully integrated into manufacturing systems, a number of things must happen. First, robot databases must be independent of the particular robot (subject to obvious limitations such as work volume, etc.). Secondly, the actual interfaces among the systems must be worked out. Thirdly, the robotic systems must be sufficiently accurate to carry out the workstation task. Robotic systems can be designed for more accuracy either by making their structures more rigid and precise, or by introducing sensors to compensate for structural inaccuracies.

Languages and control systems such as RAIL have taken an important step towards integrating robots into manufacturing systems by making the database independent of the particular robot and by allowing off-line programming.

In the case of RAIL the significance of this step has been underscored by a demonstration that took place in 1981. An interface between the AI32 controller, with the RAIL language, and a CAD/CAM database was established. The demonstration was an arc welding application that showed how parts could be designed in a CAD/CAM system and welded with the AI32 controller, the AID 800 robot, and the RAIL language. Improvements in the interface are required in the areas of flexibility and generality, but the demonstration illustrates the technical feasibility and importance of integrating robots with manufacturing and design systems.

Concluding remarks

Teach-by-showing and the use of off-line programming languages are two approaches to programming industrial vision systems and industrial robots. The former is easy to use and is adequate for many applications. The latter is more flexible and allows for more complete integration of robots into manufacturing systems. RAIL is a language which incorporates both approaches.

For many applications the interactive command module can be used to program the robotic application of the AI32. For more complex applications the programming language facilities of RAIL provide powerful and easy to use features. In an analogous manner, vision applications can be programmed either by showing the vision systems good and bad parts, or by writing a program to perform the inspection task.

The programming language facilities of RAIL allow interfacing robotic and vision systems to other factory databases, such as CAD/CAM systems. A demonstration of an interface between the AI32 with RAIL and a CAD/CAM system has taken place. Improvements on the interface are required in the areas of flexibility and generality.

References

[1] Reinhold, A.G. and VanderBrug G.J. 1980. Robot vision for industry: The Autovision system. *Robotics Age*, Fall: 22–28.
[2] Shimano, B. 1980. Levels of communication and programming methods. In, *Workshop on the Research Needed to Advance the State of Knowledge in Robotics*, Newport, RI, USA.
[3] Tarvin, R.L. 1981. Considerations for off-line programming a heavy duty industrial robot. In, *Proc. 10th Int. Symp. on Industrial Robots*, Milan, Italy.

STRUCTURED ROBOT LANGUAGE (handwritten)

IMPLICIT ROBOT PROGRAMMING BASED ON A HIGH-LEVEL EXPLICIT SYSTEM

C. Blume
University of Karlsruhe, West Germany

First presented at the 1st Robotics Europe Conference, 27-28 June 1984, Brussels.
Reproduced by permission of the author and Springer-Verlag.

An overview of the explicit programming system with the robot language
SRL and its interfaces to implicit programming and robot control level is
given. As a first step to an implicit system the robot database
RODABAS was implemented for a world model which contains all
needed data for the planning of robot actions.

Particularly in the field of assembly, developing a plan for executing a
complex task is quite a difficult job for experienced people. Therefore,
programming of industrial robots for assembly needs as much support as
possible from hardware and software tools. While the teach-in method has
been used for years in industrial applications, textual programming is now
beginning to have widespread use. Most of these programming languages for
industrial robots are of lower level and the so-called explicit languages. This
means that all robot movements and all required positions and orientations
are specified explicitly by the programmer. The future goal is the
development of an implicit robot programming system where the
programmer specifies a complete task, and the system generates all needed
robot actions and data. The base of these new implicit systems is a powerful
explicit programming 'kernel' which provides all facilities for robot control.

General requirements for robot programming languages

For the automation of assembly the robot system must be equipped with
sensors and interfaces to other machines or tools. Therefore, the robot
language should include facilities to define data structures and input/output
actions with sensors and digital/analogue interfaces.

One of the essential features of a computer language for a robot is
programming of the trajectory. For this purpose, it is necessary that the
programmer has the ability or facility to enter the start and end positions of a
movement and of the trajectory. In general, it is possible to describe

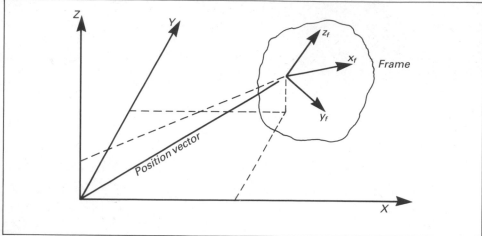

Fig. 1   Geometric position of a frame

textually any trajectory of a given coordinate system with the aid of a sequence of points. However, it is very difficult for a human operator to visualise this series of points in three-dimensional space and to describe the coordinates of each point via a programming language. In order to obtain accurate parameters, the exact location of a point is found using a measurement aid. This, however, is very time consuming and awkward. In practice, the problem is solved by leading the robot's effector through its desired path and by reading the coordinates of the corresponding arm joints at predetermined points along the trajectory. The parameters of these points are entered into the robot memory. The robot path is then reconstructed by the compiler with the aid of an interpolation algorithm. In addition to the parameters of the trajectory, the control system of the robot must have information about the orientation of the effector. This parameter can also be entered into the program by the teach-by-showing method.

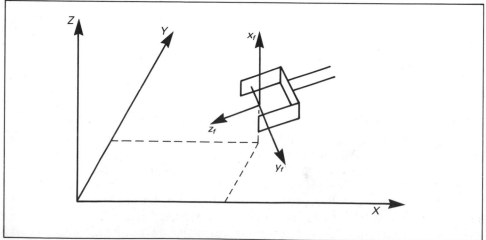

Fig. 2   Geometric position and orientation of a gripper

Modern explicit programming languages for robots use the 'frame' concept. With this method, a spatial point is described by a position vector with respect to a standard coordinate system. The orientation of the gripper is described by a rotation. Thus a frame consists of a position vector and an orientation (Fig. 1). When the effector is described by a position vector and an orientation, the following convention may be useful (Fig. 2):

- The endpoint of the position vector is located in the middle of the centre line which connects the two grippers jaws.
- The gripper points in the z-axis direction of the frame coordinate system.
- The y-axis of the effector runs through the gripping points of both jaws.

In comparing the assembly instructions for a human operator with the programming instructions of a computer language, the following points should be noted:

- Transfer of the dimensions from the drawing to the actual assembly object is performed by the operator looking at the drawing and by translating this information directly into action.
- Missing information is supplemented automatically by the operator using his experience. For example, the instruction 'assemble a flange' suffices to perform the described operations. The operator searches for the flange, places it on the assembly object and inserts the fasteners. Exact information about the position of the insertion holes is not necessary.
- The operator performs automatically a sensor-controlled positioning operation. For example, if he inserts a screw into a hole, he uses several sensors. Course positioning is done with the aid of vision and fine positioning under the guidance of the touch sensors of his fingers. In case the thread of the screw does not engage with that of the hole, he instinctively takes corrective action.
- Missing assembly element tasks may be automatically supplemented by the operator. For example, the insertion of a screw implies that a screwdriver has to be used and that the fastening process has to be done according to a given sequence of operations.
- Fixturing needed during assembly may automatically be done by the operator without explicit instructions from the drawing.

It can readily be seen that many of the above mentioned tasks have to be programmed explicitly if a robot performs the assembly. In order to obtain a fast and flexible programming system for robots, the following components should be provided:

- Installation of sensors for vision, force-torque, slip, proximity, etc., into the robot.
- Control of the robot by a run-time system which is able to adapt itself to on-line changes during assembly.
- The compiler to translate the programmed workcycle should have a component which can automatically generate missing information.

These criteria imply that the robot should be controlled by a computer and that a higher programming language is available.

Teach-in programming	Tool commands
Control structure	Parallel operation
Subroutines	Process peripherals
Nested loops	Force-torque sensors
Data types	Touch sensors
Comments	Approach sensors
Trajectory calculation	Vision systems
Effector commands	

Fig. 3   Desired features of programming languages for robots

Fig. 3 shows the desired features of programming languages for robots. In addition to the elements making up conventional languages, there should be several that are specific for robots. For example, typical data types are vector, frame, rotation and transformation. It also should be possible to describe to the robot an effector trajectory and how to handle the synchronisation of the work of several arms. The robot must be able to operate the effector and the work tools under program control. In addition, there must be language elements available that can handle signals to which the robot is capable of reacting.

Structured robot language (SRL)

Based on experience with AL (assembly language) and PASCAL, the high-level robot programming language SRL was developed at the University of Karlsruhe. The language was designed as a powerful software tool for programming complex robot assembly tasks using sensors. Therefore, much emphasis is on supporting a structured way of programming.

SRL is based on the 'frame' concept. The geometric data types, VECTOR, ROTATION and FRAME, are added and many powerful arithmetic operators are included for geometrical calculation of robot positions and trajectories. Particularly for controlling two or more robots or for the evaluation of sensor data during robot movement, multitasking facilities are provided for parallel, cyclic or time-delayed execution of program parts.

To overcome hardware dependence and to support structured and self-documenting programming, SRL includes language elements to fulfil the goals mentioned above. As a new facility, SRL has an interface to a general world model at program run-time. The world model can contain data about objects and their attributes, like workpieces, fixtures, robots, frames and trajectories.

Another fundamental element of SRL is the language PASCAL[1]. The data concept and file management is taken from PASCAL because it gives the user a very flexible and problem orientated data structure.

The goal of the development of SRL is the design of a language which can easily be learned and adapted to further developments and applications, and also to provide an interface between future planning modules and the 'traditional' programming system. A planning module will be used to generate SRL statements from a task (goal)-orientated specification instead of explicit programming of every action (see Fig. 6). Therefore, SRL has to be well-structured and universal, and has to include all features for robot programming and process control.

Concept of data

The concept of data is based on PASCAL. The standard data types INTEGER, REAL, BOOLEAN and CHAR are those from PASCAL. Also, SRL includes the structured data types ARRAY, RECORD and FILE of PASCAL. Furthermore, the programmer can define his own problem-orientated data types, as in PASCAL, by enumeration types and subrange types. Pointers are also included in SRL. As in PASCAL, the programmer can write records and its components in any expression with respect to data types.

The geometric data types VECTOR, ROTATION and FRAME are introduced as predefined RECORDS. This guarantees easy access to components of vectors, rotations or frames. A semaphor is used for the synchronisation and queuing within programs. A system flag SYSFLAG is needed for synchronisation between programs. The programmer has no direct access to the data type SEMAPHOR or SYSFLAG, but he can use the statements SIGNAL and WAIT to handle them.

Specification of system components

SRL includes a system declaration part for adopting programs to different sensors, robots and hardware facilities. With the help of the system specification, the programmer can write programs which are more hardware independent, self-documented and portable. If a program does not have a system specification the programmer has to use the predefined identifiers such as ARM or CHANNEL.

Also, ditigal and analogue ports, absolute address registers, interrupts and flags to other programs, can be specified. For access to external data from a world model or teach-in, the programmer can specify different files.

Program flow control and structure

SRL includes the block concept of ALGOL. That means in SRL a block is not only a compound statement but it can also include a declaration part. A block may contain another block and the scope rules for local and global variables are the same as in ALGOL. In SRL a different statement is included for a syntactically needed compound element to treat a sequence of statements as one.

The procedure concept allows passing parameters with call-by-value and call-by-address. If a procedure is declared with a type specification, the result of the procedure call will be a value of the specified data type (similar to the AL type procedure or PASCAL functions). Additional to the normal declarations, a procedure may contain declarations of a procedure's own variables.

These variables will hold their value from one procedure call to another but the variables are available in the procedure body only. This is helpful, for example, for programming counters.

Sections are tasks which can be executed parallel to other sections. SRL contains the traditional control structures of PASCAL:

```
–IF THEN ...ELSE ...END_IF
–CASE OF ...END_CASE
–FOR STEP  . . .TO DO ...END_FOR
–WHILE DO ...END_WHILE
–REPEAT UNTIL ...END_REPEAT
```

All statements are closed by an END_statement. To avoid undefined situations the CASE statement includes an OTHERS exit. To abort the execution of a program branch, a loop, a section, parallel parts or a program, SRL includes an EXIT statement. Program control returns after the EXIT statement to the outer control level of the program.

Move statements

To distinguish between different types of interpolation, SRL includes several move statements. Which move statements that are available depends on the implementation and the robot control. The specification of position and orientation of the tool centre point (TCP) is based on the frame concept. Reactions on sensor conditions or data are handled by the WHEN or ALWAYS WHEN statement.

PTPMOVE	Move without any synchronisation between the robot axis. Each axis is moved with maximum acceleration and speed. No general specifications are allowed.
SYNMOVE	Linear interpolation in robot joint coordinates, i.e. all axes will be synchronised. General specifications possible.
SMOVE	Movement on a straight line by linear Cartesian interpolation. General specifications allowed.
LANEMOVE	Trajectory calculation by polynoms, similar to the MOVE statement of AL. General specifications allowed.
CIRCLEMOVE	Movement along circular segment. Specifications: centre point, angular displacement, velocity or duration, fine/rough interpolation and positioning.
VIAMOVE	Move to a via frame without stopping at the via frame. The interpreter expects a next move statement for continuing the move. Only special specifications of velocity and duration are allowed.

MOVE   Move statements with interpreter or controller-dependent parameter specifications. This statement can be used for all future types of interpolation if the controller includes the control modules.

DRIVE   Movement of one or more robot axes. Specifications: velocity or duration, force, fine/rough positioning.

The general specifications for a trajectory controlled by the programmer are:

- Velocity (VELOCITY).
- Duration (DURATION).
- Acceleration (ACCELERATION).
- Constant/variable orientation during the move (CONSTORIENT, VARORIENT).
- Approach/departure points (APPROACH/DEPARTURE).
- Frames between start and end frame (VIAFRAMES).
- Force (FORCE).
- Wobble movement (WOBBLE).
- Robot arm posture (POSTURE).

Access to data defined by teach-in

With the help of the interactive component, the programmer can define frames and technological parameters by teach-in. These data are stored in a frame file. The main statements for the predefined type FRAMEFILE are:

FRAME INITIALISE (framefile) – All frames of the actual block in the program which have the same name as the frames in the internal framefile get their values from the framefile.

WRITE FRAMELIST (framefile, frames)   Adds the specified frames to the framefile or overwrites their values in the framefile.

Access to data of a world model

Much data describing the robots, machines and objects are stored in a database. This world model is not only important during planning and programming but also at run-time. Therefore, SRL includes an interface to a world model at program run-time.

The programmer can ask for the existence of an object described in the database by using the system function:

EXISTENCE (objectname)

It returns a Boolean value, reading the value of an attribute of an object:

variable := ATTRIBUTE name OF objectname ;

The program can also change a value in the world model:

> ATTRIBUTE name OF objectname := expression ;

The affixments of AL are done in the world model by the statements:

> AFFIX object1 TO object2 ; and
> UNFIX object1 TO object2 ;

The effect is exactly the same as in AL.

If the user wishes to store the actual state of the workplace environment as recorded by sensors, he can write:

> UPDATE ATTRIBUTE name OF object BY
> sensorname EVERY expression MS
> WITH expression ;

Interface to robot control

After the programmer has written the SRL program it will be compiled and translated into IRDATA-code. IRDATA is defined by the working committee of the VDI (German Engineers Association) with a large contribution from the author. After transferring the user program as an IRDATA-text to the robot control it will be executed by an IRDATA-interpreter. Therefore, SRL can be used for any robot control which is supported by an IRDATA-interpreter.

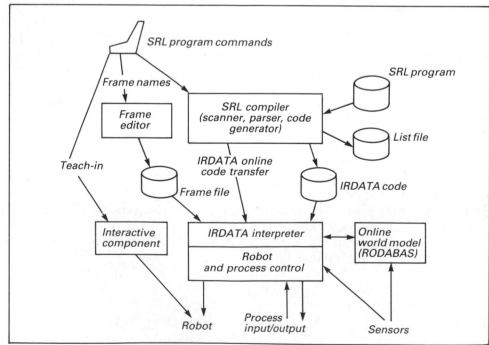

Fig. 4   SRL programming system

Programming components for SRL

The programming system (Fig. 4) includes the following major components:

- SRL-compiler.
- IRDATA-interpreter.
- Frame-editor.
- Interactive component.
- World model (optional).
- Simulator (optional).
- Symbolic debugger (optional).

The SRL-compiler reads in a SRL-program text, checks the syntax and generates an IRDATA-code similar to the P-code in PASCAL, which is executed by the IRDATA interpreter. The frame-editor and the interactive component are used to specify frames off-line or on-line by teach-in. The frames are stored in a frame list and added to the program at run-time.

Robot data base – RODABAS

Using SRL as a high-level explicit programming language the programmer has to define every frame of robot movement and every action. This is time-consuming and can lead to errors if the programmer does not have all

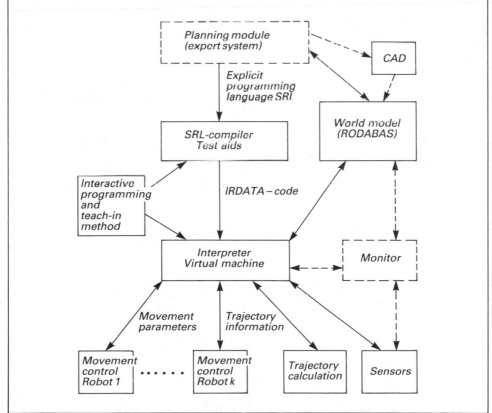

Fig. 5   Hierarchical programming structure for industrial robots

detailed information in mind. During program execution the robot control should be able to react on unforeseen events or assembly errors. Solving these problems can be done using AI methods to describe objects, structures and action sequences. Based on a special relational database, a knowledge base was implemented which includes the description of the objects to be handled, the working sphere of the robot, the magazines, the transport systems, obstacles, and other features of the robot environment.

When RODABAS is applied to systems for integrated manufacturing control, there are data of other fields like geometrical data or assembly data sheets from a CAD database. With access to this background knowledge, a planning module can generate the required robot movements and trajectories depending on the description of objects and robot environment. During run-time, with the help of sensor modules, updating can be carried out to provide data for error monitoring and error correcting (see Fig. 5).

Representation

The representation of any feature and structure requires a scheme which allows a description in such a way that it can be stored and managed by computers. The FRAME-concept of AI is based on the basic abstract features, objects, attributes and attribute values. Object means a unit which can be identified and an attribute can be related to an object and can have different values.

There is also a classification of objects with equal attributes into different types or classes. The user is able to describe in hierarchically structured

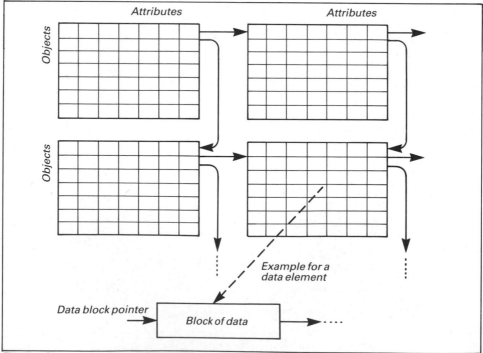

Fig. 6   *Structure of the RODABAS – implementation of the relational table*

classes or networks between them. The basic structure (object, attribute, value) is transformed directly into a relational database model. The class of objects of the same type corresponds to a relation which is represented by a table. Every row stands for an attribute; the table itself includes the values of the attributes.

Implementation

For defining or deleting relations (object classes), new objects and attributes, there is a need for flexibility within the simple structure of the table. The RODABAS-implementation guarantees this requirement by dividing the tables in the direction of rows and columns into blocks of fixed length which are linked together. By adding or deleting block chains any extension or reduction of objects or attributes is possible (Fig. 6). The objects and attributes are identified by names allowing fast access using hash-tables to calculate the table position.

Every attribute contains an attribute description to realise values of different types and lengths. The attributes descriptor includes the data type of the value, its maximum size, the range value and default value. The RODABAS-implementation allows more information, e.g. the unit of the value if it represents a physical dimension. From the frame concept an object description is derived which can include a marker for representative objects pertaining to a type or class with common attributes.

Data base functions

The database functions can be divided into:

- Structure manipulation (generate and delete relations, objects or attributes).
- Data manipulation (input, correction or delete values).
- Evaluation (output of single values or value arrays).
- Help and information.

The data access is supported by a comfortable cursor module which allows a systematic search under conditions in the relational tables. The dialogue interface is menu-orientated and self-exemplary. There is a help function for information about commands and their parameters.

Open-ended design of the world model

Based on the above described relational database, a 'kernel' of a world model for assembly operations was specified. A scheme of description was predefined and the user can apply it to describe his robot, sensors, move sequences and other features. Fig. 7 shows the predefined structure of a move in RODABAS, where the user has to give only the values and references. This structure of the object description should not be changed normally but if it is necessary to extend it for special purposes, e.g. to add information about the temperature of an object, the structure can be changed easily.

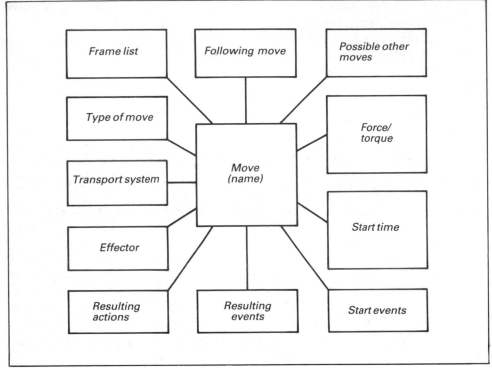

Fig. 7   Predefined structure of description of a move in RODABAS

The result, is, on one side, a finished tool for solving standard problems of the user, and on the other side, the model can be adapted or extended to any new application very flexibly.

References

[1] Blume, C. and Dillmann, R. 1981. *Free Programmable Manipulators*. Vogel-Verlag, Würzburg, West Germany. (In German)

[2] Blume, C. and Jakob, W. 1983. *Programming Languages for Industrial Robots*. Vogel-Verlag, Würzburg, West Germany. (In German)

[3] Blume, C. and Jakob, W.1983. Design of the structured robot language (SRL). In, *Proc. Advanced Software for Robotics*, Liege, Belgium.

[4] Jensen, K. and Wirth, N. 1978. *PASCAL – User Manual and Report*. Springer-Verlag, Berlin.

[5] Rembold, U. and Blume, C. 1984. Programming languages and systems for assembly robots. *Computers in Mechanical Engineering*, January 1984.

[6] D'Souza, C., Zühlke, D. and Blume, C. 1983. Aspects to achieve standardized programming interfaces for industrial robots. In, *Proc 13th Int. Symp. on Industrial Robots and Robots 7*, Vol. 1, pp.7.110-7.121. Society of Manufacturing Engineers, Dearborn, MI, USA.

[7] Müller, E. and Pods, R. 1982. *Entwurf und Implementierung einer Umweltmodell-Datenbank für die Roboter programmierung*. Diploma Thesis, University of Karlsruhe.

[8] Blume, C. 1983. Knowledge base for the future robot programming. *Elektronik* 16, August 1983.

DEVELOPMENT OF A EUROPEAN BENCHMARK FOR THE COMPARISON OF ASSEMBLY ROBOT PROGRAMMING SYSTEMS

K. Collins, A.J. Palmer and K. Rathmill
Cranfield Robotics and Automation Group, UK

First presented at the 1st Robotics Europe Conference, 27-28 June 1984, Brussels.
Reproduced by permission of the authors and Springer-Verlag.

Many robot manufacturers claim that their machines are suitable for assembly operations, but before embarking on an automated assembly project, a user would be well advised to look beyond the mechanical performance of the rival machines and to look closely at the software features offered. It is not intended in this paper to compare the mechanical features of any robots but to examine the software and control features that apply to the assembly process. In order that dissimilar devices could be compared in an unbiased manner, a benchmark specimen was designed which incorporated the most frequently encountered assembly problems. From this benchmark a number of robot control features emerged which were either essential or helpful to the assembly process. It was then possible to look at several types of robot control systems and assess the merits and demerits of each. Other factors were also taken into account such as the time taken and the degree of programming expertise required to create the control program.

There has been increasing activity in the comparison of alternative robot languages within many industrial nations, not least the USA [1,2]. However, with growing interest in the possibility of commercially led defacto standards being established [3], there is clearly a pressing need for a meaningful, industrially orientated basis for establishing the merits of alternative programming systems.

Establishing a benchmark

Arising directly from discussions held within the Robotics Europe Working Group on robot programming languages [4], steps were taken to establish a common benchmark test-piece for use by all member countries of the

Fig. 1   Complete assembly used for the benchmark

European Community. The views of experts within Robotics Europe suggested that many classes of benchmark could be conceived, however it was considered useful to make a start by developing a physical assembly which would offer a basis for joint European experience in the comparison of robot programming systems.

Characteristics of such a benchmark were to include:

- Compactness and portability – to fit inside a briefcase.
- Ability to test a variety of basic assembly operations.
- Universal application – must be capable of being assembled by the majority of assembly robots.

The benchmark assembly used for these tests is shown in Fig. 1. The components are assembled in the order indicated by Fig. 2. This test requires the robot to perform the basic assembly functions of pick, place, and insert.

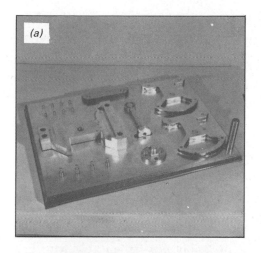

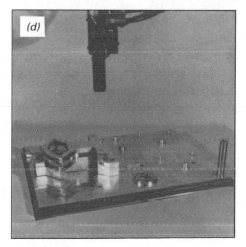

Fig. 2 Components are assembled in the order indicated

The assembly tolerances and components sizes were chosen to be within the technical specifications of the majority of assembly robots.

In order to complete this test successfully, selected systems had to be able

Table 1 Assembly related robot language and programming features

Essential	Desirable	Additional
1. Straight line motion in at least one direction	5. Straight line motion in any direction	13. Acceleration control
2. Gripper control	6. Servo control of gripper	14. Vision facility
3. Response to external signals	7. Tactile sensing	15. Software maintenance checks
4. Originate output signals	8. Step speed control	16. Control of 'settling' time
	9. Editing facility	17. CAM compatable
	10. Computational ability	18. Produces a listing if printer is interfaced
	11. Program back-up facility	19. Diagnostics
	12. Program decision making	
	20. Direct position teach facility	

to perform several basic functions. Most also offered additional features which were useful to the assembly process. Table 1 outlines the relevance of these features to assembly work.

Assembly related robot program features

It is suggested that the features shown in Table 1 perform the following functions. They either:

- Are essential for the basic assembly process.
- Improve robot assembly performance.
- Assist the programmer in making full use of the machine.

It is believed that basic assembly operations can be performed by a robot which has only one axis of straight line motion, control of an end-effector and the ability to communicate with the outside world. However, such a device would lack versatility and would probably not be suitable for very many tasks.

The range of assembly operations that may be undertaken is increased if more features are introduced into the robot's control system. One of the most important of these is individual step speed control which allows critical operations, such as pin insertion to be carried out at slow speeds and, hence, with a greater change of success.

Another feature of special interest in assembly robotics is the use of tactile sensors that enable the robot's control system to be aware if a part is not present and if excessive force is required to achieve the desired position. The conditions under which the machine took emergency actions would be preset by the robot's operating system.

Table 2   Comparison of each robot against the assembly related robot language feature

Programming feature	AML	AMLE	VAL	PAM	T³-726	IRb6
1	•	•	•	•	•	•
2	•	•	•	•	•	•
3	•	•	•	•	•	•
4	•	•	•	•	•	•
5	•		•		•	•
6	•		•			
7	•					
8	•	•	•	•	•	•
9	•	•	•	•	•	•
10	•	•	•		•	
11	•	•	•			
12	•	•	•	•	•	•
20		•	•	•	•	•
13	•					
14	•		•			
15	•		•			
16	•					
17	•		•			
18	•	•	•			
19	•	•	•		•	

Table 3   Basic moves in building the assembly

Step	Action	Comments
1	Move	General move to position above part
2	Move	Straight line move to position gripper beside part
3	Close gripper	
4	Move	Straight line motion to withdraw part from location
5	Move	General move to position part above the appropriate place
6	Move	Straight line motion to insert part
7	Open gripper	
8	Move	Straight line motion to withdraw gripper from part

Other features that would influence the performance of the robot are acceleration and 'settle' time.

The points mentioned so far apply to the physical operating conditions of the robot. However, the programming system could also be used to perform unseen tasks of benefit to the efficient use of the machine such as diagnostics, computational ability, software maintenance, CAM compatability and program back-up facilities.

Comparison of robot languages and programming techniques

Not all robots have what can be termed a 'language', but this does not mean that they cannot be programmed to perform assembly tasks. However, the complexity of these tasks would decrease with a less comprehensive programming system.

It is important to stress that this paper is not concerned with the robots themselves, only the operating systems. It was decided to cover as wide a spectrum as possible of assembly robots ranging from the most expensive to relatively inexpensive machines, all of which were from the CRAG laboratory.

Those selected for comparison were: IBM RS2 using the AML language, Unimate PUMA using the VAL language, IBM 7535 using the AMLE language, Remek PAM, Cincinatti Milacron T^{3}726, and ASEA IRb6. Those robots with no distinct language have instructions introduced via a key pad, menu or teach pendant. A comparison of each robot against the assembly related robot language feature is given in Table 2.

Table 4   Basic assembly

Step	Action	AML	AMLE	VAL	PAM	T^3-726	IRb6
1	General move	•	•	•	•	•	•
2	1, 8	•	•	•	•	•	•
3	2	•	•	•	•	•	•
4	1, 8	•	•	•	•	•	•
5	General move	•	•	•	•	•	•
6	1, 8	•	•	•	•	•	•
7	2	•	•	•	•	•	•
8	1	•	•	•	•	•	•
9	General move	•	•	•	•	•	•

Table 5   Step speed control

Step	Action	AML	AMLE	VAL	PAM	T³-726	IRb6
1	General move	●	●	●	●	●	●
2	1, 8	●	●	●	●	●	●
3	2, 7	●	●	●		●	●
4	1, 8, 7	●					
5	General move	●	●	●	●	●	●
6	1, 8, 7	●					
7	2	●	●	●	●	●	●
8	1	●	●	●	●	●	●
9	General move	●	●	●	●	●	●
10	End	●					

Robot independent, benchmark based, robot language assessment

The benchmark assembly is built up by a repetition of a number of basic moves (Table 3). This assembly can be performed by the features listed as essential in Table 1, providing that a facility exists for gripper interchange so that parts of several different sizes can be accommodated. All of the machines tested fulfilled those criteria (Table 4).

If, however, it is felt that there would be improved performance with the use of an insertion speed control facility, then AMLE may not be suitable (Table 5).

It may be believed that further improvements could be made by the use of a servo-controlled gripper. This would save the time of gripper interchange operation. If this were to be the case, then only AML and VAL would be suitable control systems.

Finally, some operations involving close tolerance insertion may endanger the gripper or the jig if too much force is used on 'stiff' parts. To overcome this, force feedback may be necessary. If this is the case then only AML of the control systems selected could be used (Table 6).

Observations

Having compared the operational characteristics of the various control systems it was felt necessary to compare the time required to program the robots. Obviously this is a difficult area in which to make objective

Table 6   Force sensing

Step	Action	AML	AMLE	VAL	PAM	T³-726	IRb6
1	General move	●	●	●	●	●	●
2	1, 8	●	●	●		●	●
3	6, 7	●					
4	1, 8, 7	●					
5	General move	●	●	●	●	●	●
6	1, 8, 7	●					
7	6	●		●			
8	1	●	●	●	●	●	●
9	General move	●	●	●	●	●	●
10	End	●					

Table 7   Time required to program the robots

Control system	AML	VAL	AMLE	PAM	T³-726	IRb6
Time taken (hours)	10	8	7	5	5	5

judgements, as different levels of programmer expertise could distort the time taken significantly. However, the data given in Table 7 is believed to be representative.

AML took significantly longer to program than the other systems. This is because of the comprehensive nature of the language and the fact that coordinates have to be entered manually because there is no 'teach' button as there is on most of the rival systems.

VAL, another comprehensive language, took nearly as long, whereas the remainder all took around five hours. It should be borne in mind, however, that these systems without the facility of a servo-controlled gripper required intermediate points for gripper changes, meaning that there was not a point-for-point comparison but one on a task-for-task basis.

Another aspect that should be considered is the ease with which a particular control system can be learnt.

To make full use of its potential, AML would require a high degree of programming expertise and a good knowledge of mathematics. VAL, on the other hand, is more user-friendly and would require less highly trained personnel.

It may well be the case that a first time user of the RS2 robot would be advised to allow a subcontractor, such as IBM, to write the software for them.

The other robots are not taught through a language and could easily be programmed by an operator after only a week or so of training by the robot manufacturer.

In general, the whole activity was felt to have been very rewarding and thought provoking. Clearly there is much more evaluation to do on the physical benchmark itself and it is now intended to refine the test-piece and circulate the benchmark around the member countries, thereby achieving a common basis for international cooperation in this field.

The implications of the benchmark are numerous. Being intentionally limited to Cartesian motions in X, Y, Z, and rotation only around the Z-axis, the benchmark clearly tests the Olivetti Sigma robot rather more comprehensively than, say, the PUMA or the IBM RS2. There is therefore an implicit problem of some importance in this direction, since benchmarks requiring more complex manipulator motions would leave a significant proportion of assembly robots unable to participate in the comparison.

There can be little doubt that as assembly benchmark work progresses it will be necessary to take steps to eliminate, as far as possible, the limiting factors associated with diverse examples of robot hardware. Conversely, it would be valuable to see some work carried out, preferably using a suitable benchmark, which compares the performance of a robot programming language, such as LM or AML, when used in conjunction with a representative range of manipulators.

References

[1] Gruver, W.A. et al. 1983. Evaluation of commercially available robot programming languages. In, *Proc. 13th Int. Symp. on Industrial Robots*, 17-21 April 1983, Chicago, Vol. 2, pp.12.58-12.68. Society of Manufacturing Engineers, Dearborn, MI, USA.
[2] Bonner, S. and Shin, K.G. 1982. A comparative study of robot languages. *Computer*, December 1982.
[3] *CAM-I Robotics Software Project*, 1983. PR-82-ASPP-01.1 CAM-I Inc., Arlington, TX, USA.
[4] *Benchmark Design*, Internal Paper. Robotics Europe – Working Group on Robot Programming, December 1983.

5

Sensory Systems

Many assembly tasks become impossible or impractical if the robot does not possess sufficient information about component orientation or the forces being exerted during assembly operations. To be aware of misalignments for example, and thereby generate an appropriate adaptive response the robot is reliant on sensory feedback. In this chapter vision and force sensors are discussed in a manner which is intended to put into context the industrial need and some representative developments in the field.

FORCE CONTROL SCHEMES FOR ROBOT ASSEMBLY

J. Simons and H. Van Brussel
Katholieke Universiteit Leuven, Belgium

Besides vision systems, force sensors have great potential for improving the autonomy and flexibility of robots in assembly systems. Although force sensors are sufficiently developed in order to be reliably used in industrial environments, the control aspects involving force feedback are not yet fully understood. The aim of the paper is to assess and to compare some potentially useful schemes involving force control. Some application examples are also given.

Up to now, automation of the assembly process has not yet undergone the spectacular technological changes seen in other engineering disciplines. Assembly is still a very work-intensive process, absorbing a large share of the direct labour costs in production engineering.

First generation industrial robots are finding widespread use in applications where the environment is completely structured and invariable. However, this latter requirement severely limits the application range for robots in assembly processes. Manufacturing tolerances make every assembly operation unique and state feedback from the environment is required in order to cope with uncertainties which are often concerned with inaccurately positioned workpieces.

Many vision systems exist in laboratories researching into object recognition, feature extraction and visual servoing. Industrial application progresses rather slowly however. Severe fundamental problems remain to be solved, such as large image processing times, focusing, recognition of three-dimensional objects and non-ideal industrial lighting conditions.

Of equal importance for increasing robot flexibility is force and tactile feedback. Force controlled robot systems can be based upon either passive or active accommodation. Accommodation is defined as the process in which the contact forces between the parts held by the manipulator and the environment modify their relative position or motion. In passive accommodation systems the position correction is generated by the contact forces themselves, whereas in active accommodation the forces are a source of information from which the position corrections are calculated.

Passive systems are simple, easy to use and fast in operation. However they have limited flexibility. Active force accommodation systems are much more flexible since their behaviour is determined by the way the force sensing loop is closed in the controller.

Force control concepts

Force feedback applications can be divided in two types:

- Applications whereby the contact force is an essential part of the process itself (grinding, polishing, deburring, etc.). Both the position of the end effector and the force exerted by the end effector on the environment have to be controlled simultaneously.
- Applications whereby the contact forces can be used as a source of information on the actual position of the endpoint (gripper) of the robot relative to the environment. Most operations during the assembly process belong to this category. Force feedback offers an important means to enlarge the allowed region of uncertainty, thus avoiding more severe requirements on the positioning accuracy of the robot and the peripheral equipment.

The following pages give a short survey on the control concepts that can be used for both types of problems.

Type I – Force control

The basic control law for applications of Type I has been formulated [1]. In a task-related Cartesian frame two orthogonal sets of directions have to be defined; a set of directions for which a desired force can be specified and a set of directions for which a desired position can be specified. Both sets of directions have to be controlled with different criteria.

Three basic control schemes can be distinguished:

- Position inner loop, force outer loop.
- Force inner loop, position outer loop.
- Position and force as parallel loops.

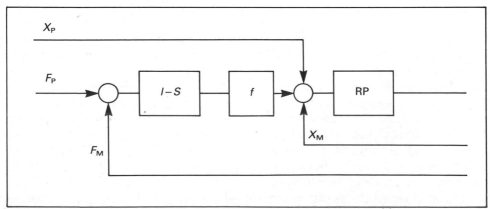

Fig. 1   *Force feedback controller described by Nevins and Whitney[2]. The force loop encloses the position loop. I–S is the selection matrix and f is the algorithm function*

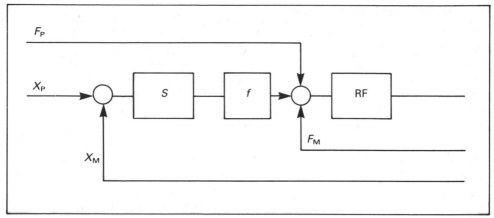

Fig. 2   Force controller described by Salisbury[3]. The position loop encloses the force loop. S is the selection matrix and f is the algorithm function

Position inner loop, force outer loop

The control scheme shown in Fig. 1 uses the oldest and best known principle of force feedback for robots. The reasons for its widespread use are historical and do not reflect the intrinsic quality of the principle itself.

The controller consists of a classical position controller around which the force loop is closed: forces and force errors are converted to position commands. The force control part consists of a selection matrix S and an algorithm function f. The selection matrix specifies the subset of force controlled axes ($s_{ii} = 1$). The function f stands for the control law that transforms force errors into position commands. In most publications, f is denoted as a matrix. Many problems, however, require a more complex control law. Also, in the formulation of f, the stability of the closed force loop has to be taken into account.

A basic objection to this type of controller is that the same controller is used for both the position and force directions (RP). Another problem is created by the incremental behaviour of the position encoders. A change in position of one encoder resolution (typically 0.1mm or more) causes very high changes in the contact forces when both the robot and the environment have normal stiffness. For this reason, some flexibility has to be built into the gripper. However, this flexibility lowers the natural frequency of the system and, by this, the bandwidth of the robot in the positioning mode.

Force inner loop, position outer loop

The problem caused by the limited resolution of the position encoders can be circumvented by closing the position loop around the force loop: positions and position errors are converted into force commands[3]. This type of controller (Fig. 2) has the same drawback as the previous one, i.e. there is only one controller (RF) designed for exerting forces on stationary environments.

Position and force as parallel loops

There are very few examples of control schemes that are based explicitly on the division of the Cartesian space in two orthogonal subsets. The basic problem is the transformation of the task space to the joint space: the subsets after transformation are no longer orthogonal, such that every axis of the robot influences the force vector and the position of the endpoint simultaneously. The free-joint method [4] ignores this non-orthogonality at first and during every sampling instant divides the robot axes into axes controlled in position mode and axes controlled in force mode. The errors in force and position resulting from this approximation are taken into account at the next sampling instant. In the hybrid force/position controller [5] every axis contributes both to the control of the force and the position. The controller (Fig. 3) has a completely parallel structure with a force control law (RF) for the force directions and a position control law (RP) for the position directions.

The selection matrix, defined in the Cartesian task space, defines the position controlled and the force controlled directions. The exact formulation of the problem in terms of the selection matrix S is of tantamount importance for the proper functioning of the scheme shown in Fig. 3. For real environments with friction in the position directions and surfaces with finite stiffness in the force directions, the parallel controller has to be completed with feedforward terms based on *a priori* knowledge of the environment (Z_o), (Fig. 4).

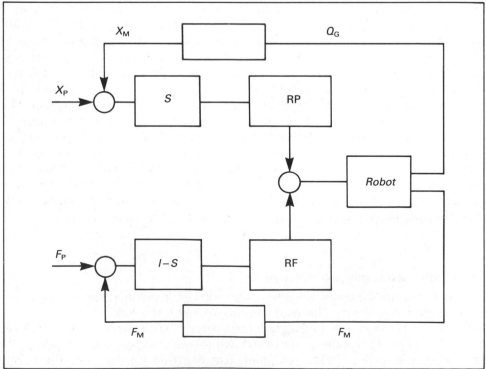

Fig. 3   Hybrid force/position controller described by Craig[5] with two parallel loops

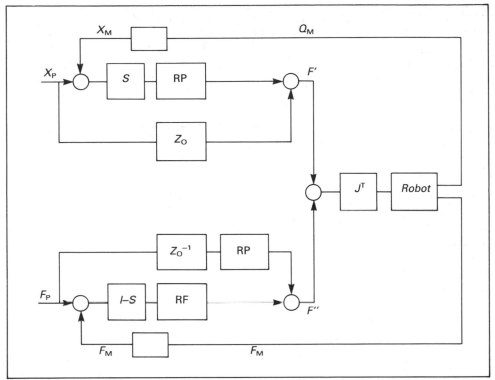

Fig. 4   Improved hybrid force/position controller. A priori information concerning the displacements in the force direction and concerning the forces to be exerted in the position direction are taken into account by a feedforward loop

Application

Fig. 5 shows a contour-following algorithm worked out for a 'position inner loop, force outer loop' robot controller[6]. An arbitrary unknown two-dimensional contour is followed with a programmed speed and contact force.

The force/position matrix S, defined in the object frame, selects position control along the surface tangent and force control perpendicular to the surface. The resulting position command is converted from the object frame to the fixed reference frame. The object frame itself is calculated from the position of the robot, the direction of the measured force, and an estimate correction for friction.

Type II – Force control

For applications when the force is not a goal to realise but an aid in positioning parts, the most successful technique is position control with controlled end-point impedance. End-point impedance can be defined as the ratio of the contact force to the difference between the programmed position of the end point and the real position of the end point. The total impedance is a cascade of three types of impedance (Fig. 6):

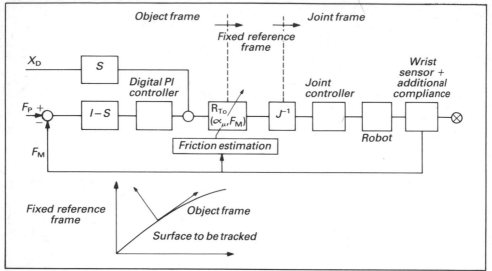

Fig. 5   Force control scheme for contour-tracking application

- Impedance of the mechanical structure.

$$F = Z_R (X_R - X_C)$$

where F is the force exerted at the end point, X_R is the real position of the end point and X_C is the position of the end point calculated from the measured joint encoders.

- Impedance of the servo controllers.

$$F = Z_S (X_C - X_S)$$

where X_S is the input to the servo controller.

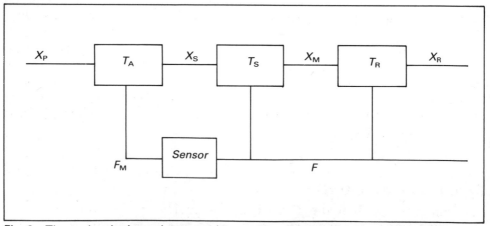

Fig. 6   The end-point impedance can be considered as a series connection of three impedances. T_A is the transmittance between the programmed position and the position modified through active force feedback; T_S is the transmittance between the modified position presented to the servo controllers and the position measured by the joint encoders; and T_R is the transmittance between the position determined via the joint encoders and the real end-point position

- Impedance as a result of active force feedback.

$$F = Z_A (X_S - X_P)$$

where X_P is the programmed end-point position.

The total impedance at the end point (Z_E) is given by:

$$Z_E^{-1} = Z_A^{-1} + Z_S^{-1} + Z_R^{-1}$$

An impedance at the end position with known characteristics facilitates the prediction of the contact forces and, if the impedance is optimised for the problem at hand, the contact forces themselves will provoke a desired correction in position. This technique, whereby the contact forces between the parts held by the manipulator and the environment modify their relative position, is called accommodation. Depending on the type of impedance that is relied upon, three types of accommodation can be distinguished:

- Purely passive accommodation (based on Z_R).
- Adaptive or programmable passive accommodation (based on Z_S).
- Active accommodation (based on Z_A).

If restricted to real impedances, the relation between the real position X_R of the end point and the programmed position X_P is given by:

$$X_R = X_P + C F$$

where C is the flexibility matrix of the end point of the robot. An accommodation system has to be designed such that for the application in mind the term $C F$ adds the proper correction to the programmed path X_P.

Purely passive accommodation

Passive accommodation is a frequently used and often unconsciously applied technique for tackling the problems caused by the uncertainties of the manipulator in relation to its working environment. In nearly every robot application the proper functioning of the system depends to a more or less high degree on a compliance in the setup. Two kinds of passive accommodation can be distinguished:

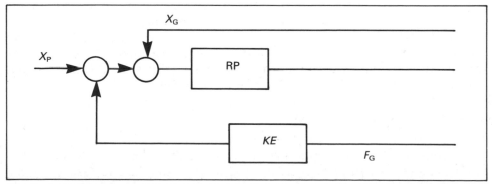

Fig. 7   In active accommodation systems, all axes are usually controlled in position mode

- Passive accommodation due to inherently present compliances can only be applied under restrictive conditions. The main objection to this type of accommodation is that the end-point impedance is not exactly known and introduces unreliabilities. A second drawback is that the allowable error window is very small. Errors larger than a few tenths of a millimetre will result in unacceptably high contact forces.

- Specially designed passive compliant structures. A lot of ingenuity has been put into the design of fixtures, grippers and tools with a compliance matrix optimised for one particular problem. One of the most interesting examples in this class of special purpose tools is the Remote Center Compliance (RCC) developed by the Charles Stark Draper Laboratories[7]. This gripper has a built-in compliance with a flexibility matrix optimised for the insertion of cylindrical bevelled shafts into chamfered holes. The most salient features of the RCC or comparable structures are their ease of application and speed of operation. On the other hand they remain a special purpose tool with an inherent lack of adaptability.

Active accommodation

The difference between active and passive accommodation is in many respects comparable to the difference between universally programmable manipulators and fixed automata. Active accommodation systems have a high degree of flexibility since their behaviour is determined by the way the force sensing loop is closed in the controller (Fig. 7). A second benefit of active systems is that the allowed errors or the regions of uncertainty are much larger since the original path of the manipulator can be changed on-line. These two elements, an easy changeover to another task and a higher adaptability, have to be weighed against the increase in complexity of a robot with active force feedback.

In trying to use the scheme as shown in Fig. 7, one has to deal with the problem of closing a force loop around a position loop. Good stability is only possible if the contact force is exerted by an ideal source such as another robot in a force-controlled mode. In real applications the contact force stems from contact with a fixed stationary environment. If the programmed end-point compliance is greater than the compliance of the environment, which is usually the case, contact will be lost due to the limitation in bandwidth of the force feedback loop. The resulting discontinuity in the measured force vector is very destabilising. The instability is further increased by the limited resolution of the positioning system (limit cycles). Therefore it is customary to add some passive compliance.

Adaptable passive compliance

This method, which is based on a programmable control of the dc-gain of the servo controllers, tries to combine the universality of active accommodation and the bandwidth of purely passive accommodation [8]. The basic principle is illustrated in Fig. 8(a) for a controller operating in Cartesian space; the normal position controller (RP) can be switched off and replaced by a linear

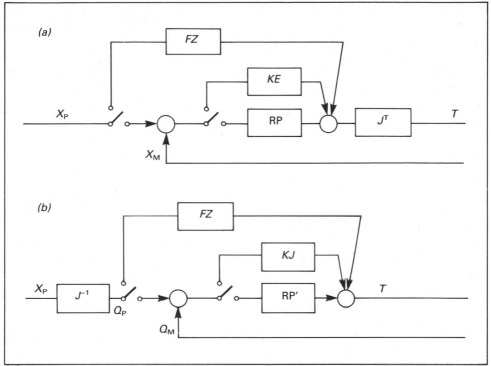

Fig. 8 Adaptive passive accommodation by (a) adjustment of the servo controllers represented in the Cartesian space, and (b) adjustment of the controllers in the joint space

matrix relation between the position errors and the Cartesian force command to the robot.

Here, KE represents the desired end-point stiffness. FZ is a feedforward term compensating the gravity forces working on the robot. Fig. 8(b) gives the same scheme for the more practical configuration of an open-loop Cartesian controller and a closed-loop joint controller. The stiffness of the joint controller KJ can be calculated from the desired Cartesian stiffness KE:

$$KJ = J^{T} KE J$$

KJ depends on the instantaneous position of the robot and is not diagonal.

There is a basic difference in the causality of the spring-like behaviour of active accommodation versus adaptable passive accommodation. In the active controller shown in Fig. 7, the offset in position of the end point is caused by an additional input to the controller based on the measured contact force. In the adaptive passive controller shown in Fig. 8, the contact forces themselves directly create the offset. In principle there is no force measurement, hence the name passive accommodation. Due to this change in causality the adaptive passive accommodation controller is not subjected to the stability problems of the active controller in making contact with the environment and to the resolution problem of the position encoders. There is no need for an extra hardware compliance while the end-point impedance remains universally programmable.

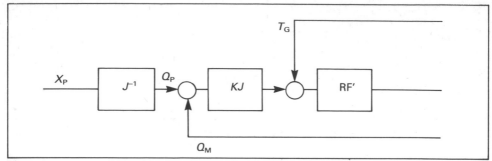

Fig. 9   *Adaptive passive accommodation with internal feedback for reducing the influence of friction in the transmission*

The scheme shown in Fig. 8 can only be implemented if the mechanical structure conforms with the principles of force/position causality, as stated above. This means that the mechanical transmissions have to be reversible and a force exerted at the end point of the robot has to result in a detectable displacement of the robot. The friction forces, reduced to the end point, have to be an order of magnitude smaller than the desired range of contact forces.

For the existing commercially available robots this is not the case. The principle of adaptive passive accommodation is perfectly suitable for wrists, multi-finger hands and directly driven robots. If a direct drive system is not feasible due to torque limitations, a quasi-reversibility can be obtained by means of an additional feedback in every axis controller from the torque measured after the transmission (Fig. 9).

Application – Precise assembly with adaptive passive accommodation

How the principle of end-point impedance can facilitate a task formulation is now illustrated [9]. Fig. 10 gives the flowchart of the insert operation of a cylindrical peg into a hole with very close tolerances. This experiment has been carried out with a five degrees of freedom Cartesian robot with reversible drives and programmable dc gains. The approach phase is rather simple in the case of bevelled edges. Care has to be taken not to convert the lateral positioning error into an angular error. This can be done by programming the stiffness as indicated in the flowchart. The same sliding behaviour can be obtained by means of a simple negative feedback matrix (active accommodation). Contact forces can be kept much lower in this case. For non-bevelled parts only active approach algorithms are possible [10].

The actual insertion is a much more severe problem especially in the case of jamming caused, for example, by surface roughness. In the second part of the flowchart a directional search algorithm has been used to find the angle of misalignment between the peg and hole. If the disturbing influence of the torques in the rotational degrees of freedom is eliminated by programming the stiffnesses as indicated in Fig. 10, the ratio of the Y to X force component gives an indication of the angle of misalignment. The higher the level of the lateral force, the more reliable this indication will be. On the other hand, it is

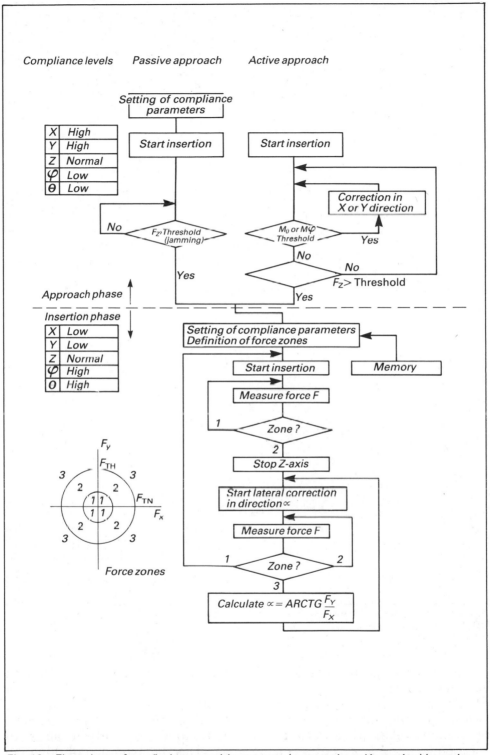

Fig. 10   Flow chart of peg/hole assembly strategy (approach and insertion) based on adaptive passive accomodation

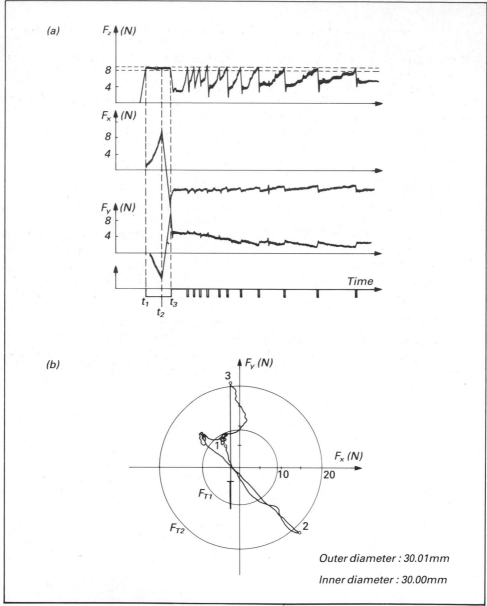

Fig. 11 Evolution of the measured force vector during assembly, in the time domain (a) and in the force space (b)

not necessary to correct exactly in the direction of misalignment in order to realign the peg.

For these reasons the force space has been divided into three zones: the zero force zone, the inner zone or the zone of uncertainty where there is an ambiguous relation between the lateral force vector and the angle of misalignment, and the outer zone. Fig. 11 gives the evolution of the force during a typical insertion both in the time domain and in the force space.

References

[1] Mason, M. 1979. *Compliance and Force Control for Computer Controlled Manipulators.* MIT Artificial Intelligence Lab., Memo 515.

[2] Nevins, J.L. and Whitney D.E. 1973. The force vector assembler concept. On theory and practice of robots and manipulators. In, *Proc. 1st CISM – IFTOMM Symp.,* pp. 273-288. Elsevier, Amsterdam.

[3] Salisbury, J. 1980. Active stiffness control of a manipulator. In, *Proc. IEEE Conf. on Decision and Control.* IEEE, New York.

[4] Paul, R. and Shimano, B. 1976. Compliance and control. In, *IEEE Joint Automatic Control Conference*, pp. 694-699. IEEE, New York.

[5] Raibert, M. and Craig, J. 1982. Hybrid position force control of manipulators. *Trans. ASME, Dynamic Systems, Measurement and Control*, 102(June): 126-132.

[6] Van Brussel, H., Simons, J. and De Schutter, J. 1982. An intelligent force controlled robot. *Ann. CIRP,* 31(1): 391-396.

[7] Drake, S., Watson, P. and Simunovic, S. 1977. High speed robot assembly of precision parts using compliance in lieu of sensory feedback. In, *Proc. 7th Int. Symp. on Industrial Robots*, pp. 87-97. JIRA, Tokyo.

[8] Simons, J. 1980. *Force Feedback in Robot Assembly using an Active Wrist with Adaptable Compliance.* PhD Thesis, Katholieke Universiteit Leuven, Belgium.

[9] Van Brussel, H. and Simons, J. 1981. Adaptive assembly. *Revue M,* 27(2).

[10] Van Brussel, H. and Simons, J. 1979. The adaptable compliance concept and its use for automatic assembly by active force feedback accommodation. In, *Proc. 9th. Int. Symp. on Industrial Robots*, pp. 167-181. SME, Dearborn, MI, USA.

ASSEMBLY ROBOT SYSTEM WITH FORCE SENSOR AND 3-D VISION SENSOR

M. Takahashi, K. Sugimoto, S. Mori and S. Hata
PERL (Hitachi Ltd), Japan

First presented at the 6th International Conference on Assembly Automation, 14-17 May 1985, Birmingham, UK. Reproduced by permission of the authors and IFS (Conferences) Ltd.

An assembly robot system which can handle unpositioned workpieces has been developed. Conventional robot and vision systems can only handle objects which are placed on a predetermined surface such as a flat conveyor belt. The developed system features a three-dimensional vision sensor, which uses two slit-light projectors and one MOS camera, and a six-axis force sensor. The system can pick up a small connector which does not have a fixed three-dimensional position and mount it correctly on a printed circuit board. The main features of an assembly robot system, which incorporates these sensors and a control language FA-BASIC, are described.

Most robots used in industrial fields are operated in a teaching-playback mode, which is generated by manipulating the robot arm under human operator control. These simple teaching-playback robots are useful in spot-welding, spray-painting or materials handling. But applications in assembly are limited only to those mechanical parts that have fixed or predetermined positions and/or orientations.

The main difference between assembly and spot-welding or spray-painting tasks is that the latter's end-effector passes on workpieces no matter where or how the workpieces are placed, whereas the former requires adjustment of the end effector's motion in relation to the workpieces' condition. The assembly robot needs sensory and language capabilities that not only control the robot motion by the teaching data, but also by visual and tactile data.

An experimental assembly robot system consisting of four robots with sensory and language capabilities has been developed. Using this system, the performance of each individually developed component was verified by assembling a typical electromechanical product, in this case, a floppy disk drive.

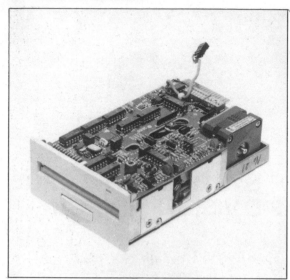

Fig. 1   Subassembled floppy disk drive

Typical assembly process

The main components of the floppy disk drive are the chassis, including motor and mechanical parts, and the printed circuit board (pcb). Assembling the mechanical parts onto the chassis is automated by the conventional playback method, while the assembling of pcbs is done by an automatic insertion machine or an automatic parts mounting machine.

Although this subassembly is performed automatically, the final assembly requires some difficult tasks such as connecting wires between the pcb and the motor.

Fig. 1 shows the subassembled floppy disk drive, and Fig. 2 shows the typical assembly process of this type of electromechanical product.

Assembly line

Fig. 3 shows the configuration of the system. The assembly line consists of four robots, three image processors and two six-axis force sensors. Each robot incorporates the assembly process shown in Fig. 2.

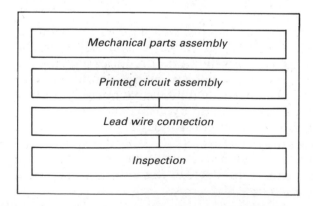

Fig. 2   Typical assembly process of electromechanical products

Fig. 3 Robot line for assembly

Virtual compliance control

Two types of chassis are fed from the left-hand side by a belt conveyor which does not have any mechanical fixture for positioning the chassis. The type, position, and orientation of the chassis are identified by an image processor

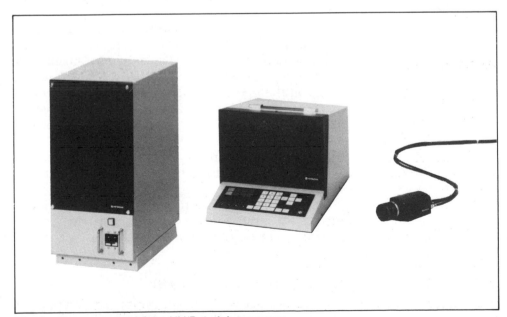

Fig. 4 Components of the HV/R-1 vision system

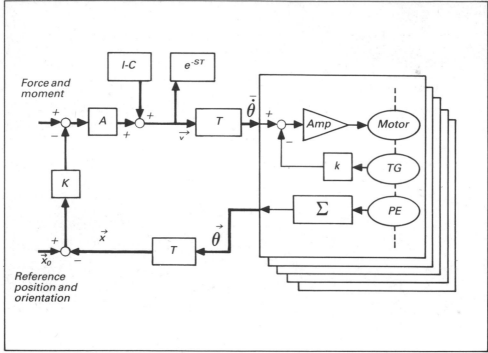

Fig. 5 Block diagram of the robot control. T is the coordinate transformation matrix, K is the stiffness coefficient matrix, A is the force – velocity transformation matrix, and C is the damping coefficient matrix

using a TV camera located above the conveyor belt. This image processor, the newly developed HV/R-1[2] shown in Fig. 4, is specially designed for use and coordination with robots.

Teaching the position and orientation is performed by a teaching-by-showing method. In the first stage, the robot carries a reference chassis to a position on the conveyor belt under the camera. The camera transmits the chassis image to the image processor which memorises the position and orientation data as x_0, y_0, θ_0. In the operating mode, the image processor's output is $\triangle_x$, $\triangle_y$, $\triangle_\theta$, which represents the deviation of the actual position and orientation of the chassis. The robot modifies its end-effector position using this data.

A second conveyor belt transfers the chassis between the robots. The chassis are fixed on a pallet on the conveyor belt using force-controlled mode.

Fig. 5 is a block diagram of the robot control. The control method[3], 'virtual compliance control', uses the force resulting from the contact between the chassis and fixture to move the robot until the chassis is seated in the fixture.

A control language for factory automation, FA-BASIC

For simple repetitive work, it is not necessary to use symbols or language to describe the given task. But to perform an assembly task which involves

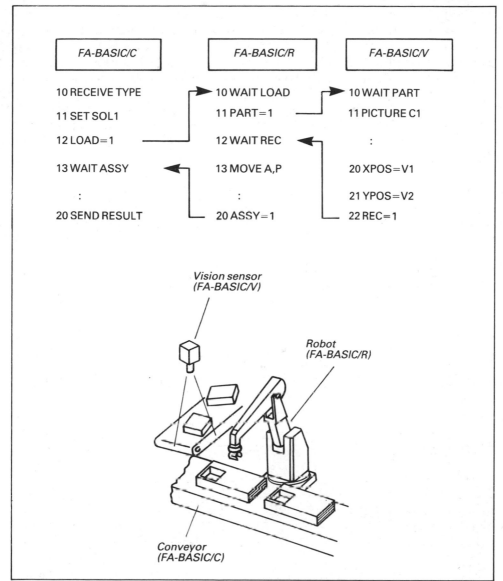

Fig. 6 Examples of FA-BASIC's description. FA-BASIC is a unified language for robots, vision sensors and programmable controllers. It consists of problem orientated commands and common BASIC commands

several or various components, or to modify a predetermined motion by the signals of sensors, it is necessary to have language capability. By describing the component's control by a unified control language, not only can the amount of man-hours for developing the system be reduced drastically but also a more sophisticated control is made possible.

For this purpose, a unified control language for factory automation, FA-BASIC, has been developed. Fig. 6 describes how the commands of FA-BASIC are used in controlling a system.

Handling unpositioned connectors

The handling of unpositioned connectors is often a difficult task, since the position and orientation of workpieces at the end of a wire is not only two-dimensionally uncertain, but also three-dimensionally uncertain. Therefore, to handle this type of workpiece, 3-D image-processing technology is required.

The 3-D handling robot developed consists of a robot arm with six-degrees of freedom of motion, and a 3-D image sensor. The 3-D image sensor is composed of two slit-light projectors and one MOS camera. By using a calibrated range map, it is able to get the range data of the object from the image of the workpiece taken under the projected slit of light. The system also has a six-axis force sensor on the wrist, which detects forces and moments.

To determine the position of the object, the robot initially traces the lead wire using the 3-D image sensor. Once it finds the object it grips it from the normal vector direction and inserts the gripped connector into the female connector under the force control mode. Fig. 7 shows the configuration and external view of the system.

Inspecting the assembled work

As indicated in Fig. 2 inspection usually follows assembly. Visual inspection, particularly of pcbs, is certainly required, but the TV camera image of a pcb is so complex for an image processor that it is often difficult to distinguish correctly and incorrectly assembled products.

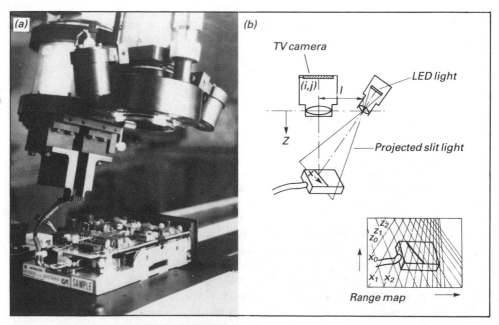

Fig. 7   Handling of unpositioned connector: (a) external view of the 3-D vision sensor and (b) principle of the 3-D vision sensor

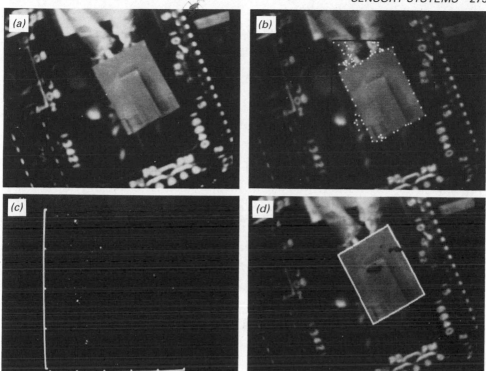

Fig. 8   Inspecting the assembled printed circuit board: (a) original image transmitted through the TV camera, (b) extracted contour by two-state thresholding and polygonal approximation, (c) noise elimination by Hough transformation, and (d) the obtained four sides of the connector

To solve this problem, a fast 'Hough transformation' is applied to the image processing. The method is a kind of noise elimination technique, and is explained as follows using the example of defect detection in connector assembly:

- Thresholding and polygonal approximation. Fig. 8(a) shows the original image transmitted through the camera. The image is stored in a $256 \times 256 \times 8$-bit grey-scale memory. A reference point in the zone where the connector is supposed to exist is then chosen, and the image is thresholded by the reference point's brightness $\pm \epsilon$. The dots in Fig. 8(b) show the contour of the obtained image.
- Since the shape of the object is rectangular, the Hough transformation is applied. Fig. 8(c) shows the transformed points in the r-Θ plain. Four clusters of points in the figure correspond to the four sides of the connector. Fig. 8(d) shows the result of processing.

Concluding remarks

In summary, an assembly robot system which can handle unpositioned workpieces was developed. In achieving the success of the system, the following elements were developed:

- HV/R-1 image processor.
- Virtual compliance control technology.
- Control language for factory automation, FA-BASIC.
- 3-D handling robot.
- Image processing technology for recognising objects from a complex scene.

In applying the system for floppy disk drive assembly, the usefulness of these developments were verified.

References

[1] Takahashi, M. 1984. Development of fundamental technology for assembly automation. *Hitachi Hyoron*, 66 (10): 35-40.
[2] Hata, S. 1984. A compact image processor HV/R-1 for robot. *Hitachi Hyoron*, 66 (10): 21-26.
[3] Hirabayashi, H. 1984. Force feedback control for multi-degree of freedom robot. In, *27th Jido Seigyo Rengo Koenkai*, pp. 241-242.

SENSORY CONTROLLED ROBOT ASSEMBLY AND INSPECTION STATION

J. Spaa, N.J. Zimmerman and U.G.G. Singe
Delft University of Technology, The Netherlands

First presented at the 14th International Symposium on Industrial Robots and 7th International Conference on Industrial Robot Technology, 2-4 October 1984, Gothenburg, Sweden. Reproduced by permission of the authors and
IFS (Conferences) Ltd.

In accordance with the Dutch research programme, Flexible Production Automation and Industrial Robots (FLAIR), Delft University of Technology has started the development of a flexible assembly cell. This paper presents the definition and description of the hardware set-up of a development station for research on sensory controlled robot assembly and inspection in the Laboratory of Applied Physics. This station is being equipped with vision for the inspection and identification of parts and the measurement of position and orientation, and with force and tactile sensing for assistance in assembly tasks. The visual inspection of the final product is provided for. Results on the flexible assembly of parts for gas water-heaters are presented.

In recent years, the Dutch government has stimulated the research on and the industrial application of flexible automation and industrial robots in programmes called 'FLAIR' and 'Demonstration projects flexible automation'[1]. The funds available for these programmes amount to 2 and 18 million Guilders, respectively. Within the framework of the research programme 'FLAIR' the Departments of Mechanical Engineering and Applied Physics of Delft University of Technology have set up a joint project 'Flexible Assembly Cell'.

The Department of Mechanical Engineering is developing an assembly cell for pneumatic break cylinders with special emphasis on research on cell control systems. The Department of Applied Physics is constructing a sensory controlled robot assembly and inspection station and will concentrate on research in the field of robotic sensory control and inspection. The station is tailored to the assembly of hydraulic lift assemblies – parts of a gas water-heater.

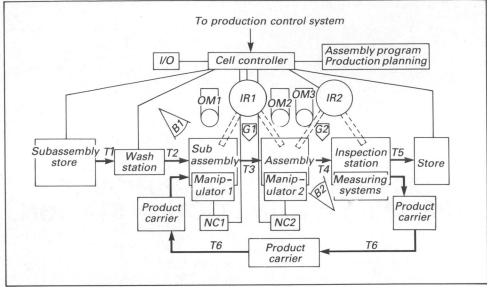

Fig. 1   Structure diagram of a flexible assembly cell, equipped with advanced control and sensor systems. IR1 and IR2, industrial robots; T1 – T6, transport devices; OM1 – OM3, part feeders; B1 and B2, vision systems; G1 and G2, tool magazines; and NC1 and NC2, manipulator controllers

In time, the experience gained in both projects will be used for the construction of a flexible assembly cell in which two industrial robots will operate (Fig. 1).

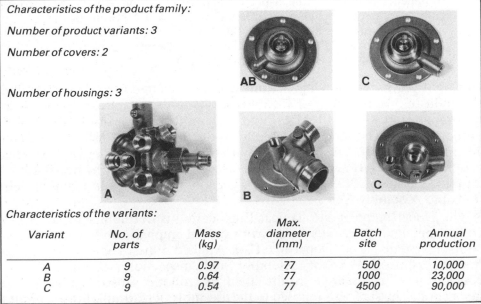

Characteristics of the product family:

Number of product variants: 3

Number of covers: 2

Number of housings: 3

Characteristics of the variants:

Variant	No. of parts	Mass (kg)	Max. diameter (mm)	Batch site	Annual production
A	9	0.97	77	500	10,000
B	9	0.64	77	1000	23,000
C	9	0.54	77	4500	90,000

Fig. 2   Product family of hydraulic lift assemblies

Fig. 3 Exploded view of a hydraulic lift assembly (variant A): (a) bolts, (b) cover, (c) 0-ring, (d) diaphragm, and (e) housing

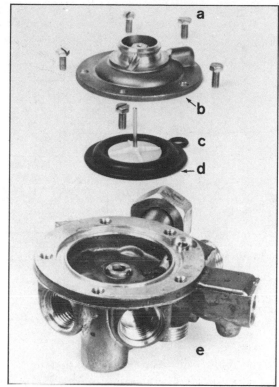

Choice of product

In realising the assembly station the choice of product is most important. It needs to be 'representative' of assembly, and assembled with the most frequently occurring functions, that of fitting one part to another and bolting.

As a result, a family of hydraulic lift assemblies for gas water-heaters (Fig. 2) of the Dutch manufacturer FASTO was chosen. This product family has the following features: three variants (one type is shown in Fig. 3), a total mass ranging from 0.5 to 1kg, and a daily production of 500 pieces per shift, making it suitable for assembly in a flexible assembly station, as research on the application of industrial robots for assembly shows[2]. In addition, because of their mass, dimensions and number of components, the hydraulic lift assemblies form a representative example of assembly products[3,4]. Also, the product does contain the most frequently occurring assembly functions and is assembled from bottom to top. Nevertheless, typical feeder problems, such as unknown orientation, product identification and feeding of different parts, were encountered.

Assembly and inspection station

The layout of the assembly and inspection station is shown in Fig. 4. The mechanical parts of this station are a transport system with two docking stations, an industrial robot, a manipulator, a pneumatically powered

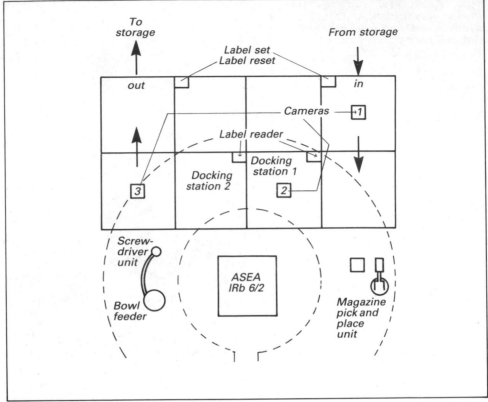

Fig. 4   Layout of the robot assembly and inspection station

screwdriver unit and a vibratory bowlfeeder. A computer vision system with four cameras forms part of the station. Three cameras (see Fig. 4) are mounted vertically above the transport system, whereas the fourth camera is mounted on the robot.

Storage and transport systems

To simulate an industrial environment, the station has been built around a transport system: the Variable Transport System (VTS) of the Philips' Nederlandse Machinefabriek 'Alkmaar'. It is a modular system; modularity applying both to the mechanical construction and to the control structure. The VTS is equipped with two 8031-based controllers, which route product carriers through the system. For that purpose they obtain information about the position of the product carriers from micro switches and proximity switches. For identification, the product carriers are provided with a simple 2-bit active label. To simplify the use of computer vision some of the carriers are provided with a transparent window. This offers the possibility to use backlighting, resulting in high-contrast images.

For handling purposes, a small manipulator with two degrees of freedom was constructed. It is pneumatically driven and moves against fixed mechanical stops. The gripper consists of two flexible fingers which bend under air pressure. The manipulator is attached to a storage system containing stacked parts. This stack can be moved vertically by means of a stepper motor. In this way, parts are presented to the manipulator. The manipulator/storage system is controlled by a simple PLC-like device.

A manually loaded vibratory bowlfeeder presents orientated bolts to a pneumatically powered screwdriver.

Docking station

Two docking stations are included in the transport system. They can clamp product carriers with an accuracy of approximately 0.1mm. One carrier is provided with a general purpose fixture designed to accommodate all three types of housings. Assembly can be performed on this carrier in one of the docking stations.

The robot

The station is equipped with an ASEA IRb 6/2 robot, which has an anthropomorphic construction with six degrees of freedom. The handling capacity is 4kg, and has a repetition accuracy of ± 0.2mm. The axes are driven by dc-motors.

In addition to the standard programming facilities the robot controller is equipped with adaptive control and a computer link. The adaptive control option enables the use of external sensors to control speed of movement, to follow contours, or to perform searching movements. Sensors can either be fixed in space or mounted on the robot. In the latter case, the controller takes into account the orientation of the robot when calculating the direction of corrective movements. The speed of the robot is controlled by a one-dimensional signal. Up to three signals can be used to guide the movements of the robot during searching and contour following. Sensor information can be presented to the controller in a 1-8 bits format or, alternatively, as an analogue signal.

The computer link enables asynchronous communication with a computer. This allows up- and down-loading of robot programs. Thus, a relatively large computer background memory is available for use as a program bank. Besides this, a computer can control and monitor program execution. Direct readout of the position of the robot, of sensors and other inputs of the controller, as well as direct control of the robot's movements, is possible.

Sensing

The role of sensors in the assembly process

Sensors can play an important role in assembly operations. They can be used to monitor and control the process, to verify tasks and to detect abnormalities, thus giving a higher reliability of the process.

Sensor systems can also provide a certain amount of adaptability. This means they enable compensation for deviations in the dimensions of parts or tools, and enable control or modification of the movements of the robot, during acquisition of incorrectly positioned parts. Also, sensor systems can provide flexibility through their ability to distinquish different parts when randomly presented. In addition to the assembly of single types of products, sensors enable assembly of different products on a single station. Another task for sensor systems is in-line inspection of parts and subassemblies.

Generally, an assembly process includes the following steps: handling (presentation, acquisition, separating), composing (mating, joining or bonding) and checking. Parts can be presented in various ways. When mechanical devices are used, the positional accuracy can be high. Parts can be constrained by passive devices such as containers, fixtures, totes or bins, or by active devices such as feeders or manipulators. Due to the different shapes and dimensions which have to be accommodated, such devices cannot always be used in a flexible assembly process. This can give rise to situations in which position or orientation is not sufficiently restricted. Sensors can help to resolve this uncertainty. Either two- or three-dimensional vision systems and/or rangefinders are most suitable for this purpose, but ultrasonic ranging and, in some cases, force measurement or tactile sensing are also useful. During gripping and manipulation of parts, the gripping force and the distance between the closed fingers can provide valuable information. Tactile sensing enables detection of the way in which parts are positioned in the gripper and of the occurrence of slip. Part identification can also be performed, but is limited by the low resolution and the small number of elements of the presently available sensors.

Parts are mated subject to contact forces. Measurement of the forces and torques provide the most reliable source of information to guide the movements of the robot[5]. The determination of forces is usually carried out in an indirect way. When using electrically driven robots, the motor current can be used for this purpose. A more direct and more precise way is the measurement of the deformation of a complaint device, instrumented with strain-gauges, LVDTs or piezoelectric transducers. These measuring devices can be mounted between the robot and its end-effector, and can either be integrated into the gripper or form part of a fixture. In these cases the sensor system usually provides a multidimensional signal which for example has to be transformed to three force and three torque components. The measurements and calculations involved can be performed at such a rate that insertion of force sensing inside the feedback loop is possible[6]. Compensation for angular or lateral misalignment can be achieved using the built-in compliance or, better still, using an engineered compliance such as remote centre compliance[7]. Although this provides a reliable, fast and relatively cheap solution, flexibility is low. The combination of a force/torque sensor and a compliant device can yield a flexible, fast and reliable solution to the part-mating problem.

Besides the sophisticated sensor systems mentioned above, a variety of simple sensors are in use. They are often binary in nature and provide status information such as presence or absence and contact or no-contact.

Sensor systems of the station

The assembly and inspection station will be equipped with vision, force/torque and tactile sensing. At present, the Automatix Autovision 2 (AV2) system is being used. This will eventually be replaced by the more

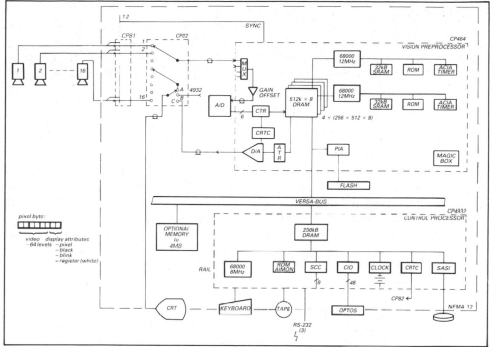

Fig. 5   Architecture of the multiprocessor, grey-scale system Autovision 4

powerful Autovision 4 (AV4) system. The architecture of the AV4 is shown in Fig. 5. Up to 16 cameras with a maximum resolution of 512 × 512 pixels can be connected. Binary as well as grey-level processing is available. Typical processing times are well below 1s. The system can be used for a variety of purposes such as the determination of position and orientation and for the performance of complex inspection tasks. It is programmed in the structural RAIL language which resembles PASCAL.

As a second vision system the Automatic Visual Inspection System (AVIS) is available[9]. It is a low-cost binary system developed at Delft University in cooperation with the Institute of Applied Physics TNO/TH.

In the Pattern Recognition and Image Processing Group of the Department of Applied Physics a cellular logic processor (CLP) has been designed and is under construction. This is a single board processor, which performs image-to-image processing.

Through a remote link there is access to the Group's image-processing system[10,11] (see Fig. 8). This comprises a number of displays, input devices, the Delft Image Processor (DIP) and a VICOM image processor. A cellular logic image processor (CLIP) will be added shortly.

Force sensing is available through an Astek FS6-120A-200 sensor (Figs. 6, 7). It can measure all six components of force and torque. Forces up to 200N are measured with a resolution of 0.1N. Torques up to 4Nm are measured with 0.002Nm resolution. The data rate is programmable from 1.8 to 240Hz. All electronics, including the processor system, are located inside the sensor housing.

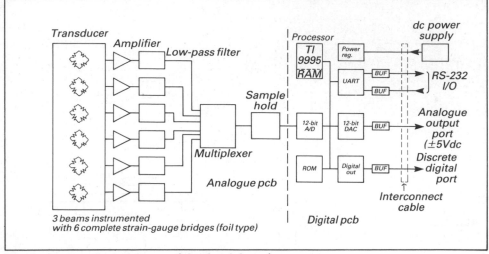

Fig. 6   Electronics hardware of the Astek force/torque sensor

In cooperation with the Institute TNO/KRI a tactile sensor system based on the piezoelectric material PVDF is under development. It will consist of approximately 16×16 elements of $1mm^2$.

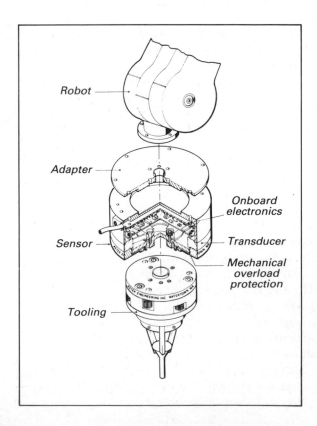

Fig. 7   The Astek force/torque sensor

System integration

Design considerations

The assembly station consists of a variety of devices. The way in which these are linked to one another greatly influences the station's performance, its flexibility and ease of programming. Some devices are not programmable, or make use of permanently stored programs. In this case, the communication concerns only simple commands or data. When programmable devices are concerned, off-line programming can be useful and sometimes even necessary. This means that up- and down-loading of programs will have to be supported. To facilitate set-up, debugging, experimentation and training, it must be possible to operate all devices separately. The designed control structure reflects these general considerations.

System description

A block diagram of the station is shown in Fig. 8. Program development can be performed on the main processor unit (MPU). This unit is a VME-bus compatible single board computer featuring a 12MHz 68010 processor, a

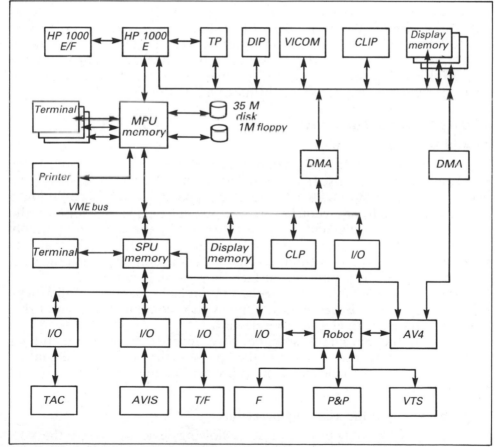

Fig. 8 Block diagram of the assembly and inspection station

1 Mbyte dual-ported 100ns RAM, two memory management units, a fast-seek 35 Mbyte Winchester disk and a 1 Mbyte floppy disk unit. It is a multi-user system running under the UNIX V operating system. Via the VME-bus, programs can be down-loaded to the AV 4 and to the station processing unit (SPU) and, via one of the SPU's RS232 lines, to the robot controller.

The SPU provides real-time performance. It connects to the sensor systems as well as to the robot. It can perform processing on sensory data and presents the data in a form which is acceptable for the adaptive control of the ASEA robot. Via the computer link, the SPU will be able to directly control the robot. The SPU contains a 10MHz 68010 processor and a 128 Kbyte dual-ported 150ns RAM. It connects to both the VME-bus and the G64-bus. The G64-bus is used to connect the SPU to a tactile sensor (TAC), the automatic visual inspection system (AVIS), the Astek torque/force sensor (T/F) and some adaptive control inputs of the robot. Other adaptive control inputs are directly connected to the AV4. Feeders (F), the pick-and-place unit (P&P) and the variable transport system (VTS) are under direct control of the robot. The addition of a cellular logic processor (CLP) and a display memory provide on-line image processing capabilities.

The devices on the VME-bus and the AV4 will be connected to the image processing system of the pattern recognition and image processing groups. This is a multi-user system[10,11] controlled by a HP1000/E. The MPU is connected with the HP over a RS232 line. Image transports are intiated by the HP but controlled by the transport processor (TP) over the 16-bit 800kHz image bus. Via a DS1000 link, the HP1000/E is coupled to the HP1000/E and F processors of the Department of Applied Physics mainly to provide access to tape units, line printer and plotter.

The hydraulic lift assembly

Assembly process analysis

The first application in the assembly station concerns the assembly of a family of hydraulic lift assemblies. For each of the product variants a specific housing and cover have to be selected, and supplemented by five bolts and a diaphragm, which are common to all variants. This necessitates the classification of housings and covers. These parts are presented in a partially positioned way, lying on a specific side on a flat surface. This means that only two coordinates of their position and the angle of rotation around the axis, perpendicular to the supporting surface, have to be determined.

A typical feeder problem is the presentation and separation of the sticky rubber diaphragms. Because each diaphragm is combined with an O-ring both position and orientation are relevant.

Fitting to the housing has to be very accurate because the rubber diaphragms will not easily slide into the groove in the housing. The housing and cover have to be held together with a certain amount of pressure and fastened with five bolts.

Implementation of the assembly procedure

In the present assembly procedure the following devices are used: the VTS transport system, product carriers with transparent windows, a carrier

Fig. 9   Display of the Autovision 2 system. Determination of position and orientation of the housing of variant A

provided with the general purpose fixtures, the ASEA IRb 6/2 robot, and the AV2 vision system with two cameras connected.

The housings and covers are transported by means of the VTS. When a carrier arrives at the first camera position, a picture is taken, then type recognition is performed: objects are classified as a carrier with fixture, or as a certain type of housing or cover.

Carriers pass until a housing is detected. This carrier is positioned at docking station 1. Using the second camera, position and orientation are determined to enable the robot to pick the housing from the carrier (Figs. 9, 10). Then, a carrier with fixture is positioned at docking station 2 and the robot places the housing in the required way into the fixture. A third carrier transports the cover to docking station 1. The robot acquires this cover in the same way as the housing. The last step is the fitting of the cover on the housing.

Results

For the identification of the carriers, the housing and the covers, a high-contrast image obtained by backlighting of the carriers is used. Segmentation is achieved by thresholding. Subsequently, the objects are

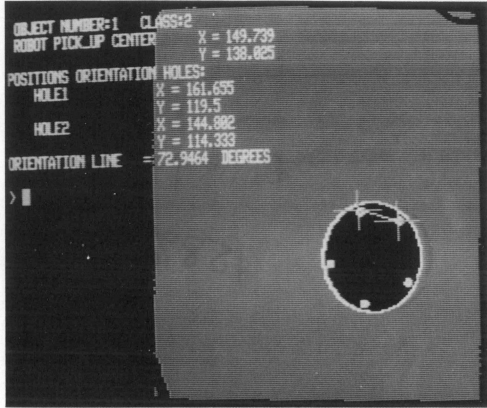

Fig. 10   Display of the Autovision 2 system. Determination of position and orientation of the cover of variant A/B

classified in a decision tree, using the following object features: area, number of holes and maximum radius from the centre of gravity to the contour. Classification is verified using circularity and the quotient of the minimum and maximum radii. Depending on the specific part, the classification time varies from 0.9 to 1.4s.

At the first docking station, vision is used to guide the robot for part acquisitioning. Position and orientation are calculated from the measured coordinates of the centres of gravity of a number of specific holes in the objects. For example, in the case of variant A, the two holes indicated in Fig. 9 and the centre hole are used. These holes are identified by their mutual distance and their distance to the object's centre of gravity. For variant A the determination of position and orientation takes 0.7s. The positional and the angular accuracy amount to ±0.4mm and ±0.04°, respectively, in a field of view of 300 × 400mm. This is more than sufficient for this application.

Values for the other covers and housings have a comparable order of magnitude. The identification of the holes in the housing of variant C, however, is more time-consuming, resulting in a processing time of 1.4s for the determination of position and orientation.

Acknowledgements

The authors would like to thank A. Kamp and A. van de Stadt for their contributions in programming the Autovision, and ASEA and F.F. van Leeuwen and W. van Oel for their technical support.

References

[1] Lafaille, A. 1983. *Strategie National d'Automatisation. Les politiques de Recherche/développement en Robotique.* CESTA, Paris, pp.150-154.

[2] Warnecke, H.J. and Walther, J. 1983. Programmable assembly station for the automatic assembly of car aggregates. In, *Proc. 15th CIRP Int. Seminar on Manufacturing Systems*, pp. 290-299. CIRP, Paris.

[3] Drexel, P. 1982. Modular flexible assembly system 'FMS' for Bosch. In, *Proc. 1st Int. Conf. on Flexible Manufacturing Systems*, pp. 171-195. IFS (Publications) Ltd, Bedford, UK.

[4] Swift, K.H. 1978. Classification for automatic assembly of small products. *Ann CIRP*, pp. 435-440.

[5] Sanderson, A.C. and Perry, G. 1983. Sensor-based robotics assembly systems: Research and application in electronic manufacturing. In, *Proc. IEEE*, 71 (7): 856-870.

[6] Hirzinger, G. 1983. Direct digital robot control using a force-torque sensor. In, *IFAC Symp. on Real-Time Digital Control Applications*, Guadalajara, Mexico.

[7] Whitney, D.E. 1982. Quasi-static assembly of compliantly supported rigid parts. *Dynamic Systems, Measurement and Control*, 104 (March): 65-77.

[8] Zimmerman, N.J., Van Boven, G.J.R. and Oosterlinck, A. 1983. Overview of industrial vision systems. In, *Industrial Applications of Image Analysis*, pp.193-231. DEB, Pijnacker, Netherlands.

[9] Zimmerman, N.J., van den Boomgaard, R. and Groen, F.C.A. 1983. Coupling of a low-cost vision system on an industrial robot. In, *Robot Vision in Holland*, pp. 195-207. DEB, Pijnacker, Netherlands.

[10] Duin, R.P.W. 1983. Software systems for image processing. In, *Industrial Applications of Image Analysis*, pp.141-152. DEB, Pijnacker, Netherlands.

[11] Duin, R.P.W. and Gerritsen, F.A. 1984. *The Delft Image Analysis Laboratory: A Multi-user Facility for Research, Development and Education of Image Analysis Methods.* Internal Report, Delft University of Technology, Netherlands.

UNPACKING AND MOUNTING TV DEFLECTION UNITS USING VISUALLY CONTROLLED ROBOTS

P. Saraga, C.V. Newcomb, P.R. Lloyd, D.R. Humphreys
and D.J. Burnett
Philips Research Laboratories, UK

First presented at the 3rd International Conference on Robot Vision and Sensory Controls, 6-10 November 1983, Cambridge, MA, USA. Reproduced by permission of the authors and IFS (Conferences) Ltd.

TV deflection units, loosely constrained in a large carton, are unpacked by a gantry robot. The robot is equipped with a TV camera and a parallel projection optical system, which are used to determine the position and orientation of the deflection units in the carton. The deflection unit is then mounted in a specific orientation, onto a picture tube by a PUMA 560 robot equipped with three TV cameras to locate the deflection unit and the neck of the tube. This paper describes the structure and operation of both systems, including grey-level picture processing, camera calibration without operator intervention, and the use of a general purpose, robot operating system, ROBOS, to control the two tasks.

The use of visually controlled robots in flexible assembly systems is becoming a practical reality. Such systems can in principle handle and assemble parts presented in a variety of attitudes, can compensate for changes in their own physical structure, and can be reconfigured to a new task.

One stage in the manufacture of a Philips television set consists of the removal of a deflection unit (DU) from the packing in which it arrives from another factory, and its correct placement on the neck of the picture tube. The systems described in this paper to perform these tasks are structurally similar to many sensor-controlled robot systems, including a number developed at Philips Research Laboratories[1-3]. Each one contains a control program which specifies the task to be performed. The control program invokes vision system commands to determine identity or to extract location information, which is then converted to the reference frame of the manipulator that performs the required mechanical actions. There are also facilities for sequencing the operation of the various parts of the system, and for coping with foreseeable errors and failures.

The success of such systems depends significantly on the use of standard mechanical and software modules, and the ease with which these modules can be reconfigured to perform a different task when required. If such systems are constructed using a single manufacturer's equipment, then a number of system building aids and programming tools are often provided. Each of our systems, however, had a different vision processor, a different manipulator, and different numbers and types of computer. To cope with such situations, a ROBot Operating System, ROBOS, is being developed. ROBOS coordinates the activities of the various machines, maintains information about the relationships of all frames of reference, transforms locations from one frame to another, supervises a system of process interlocks and generally deals with much of the 'housekeeping' necessary in a robot application.

Unpacking deflection units

The DUs are packed in three layers in a cardboard carton measuring $1 \times 1.2 \times 0.6$m high. A layer consists of fourteen boxes each containing four DUs packed with their axes of symmetry horizontal and with each DU rotated (about a vertical axis) by 180° from its neighbours. The initial misplacement of a DU relative to a grid based on the nominal carton dimensions is ±15mm, with a misorientation about a vertical axis of ±10°. This is caused by the play between the DU and its box, and between the box and the carton. The misplacement is too great to enable a blind robot to unpack the DUs. The task of the vision system is to find the position and orientation of each DU sufficiently accurately for a mechanical gripper to grasp it and withdraw it from the packing.

Mechanical structure and operation of the robot

A Cartesian gantry robot with an X, Y, θ carriage, and an arm which can move vertically, was constructed to perform the unpacking task. The arm carries the specially designed gripper for removing the DU from the carton. The TV camera is mounted on the carriage, so that it maintains the same X, Y, θ position as the gripper. There is also a roller conveyor to carry the carton of DUs, and a conveyor belt onto which the DU is placed after unpacking.

The roller conveyor moves the carton to approximately the right position within the working space of the robot. The robot then positions its gripper above the first DU (Fig. 1), and a picture is acquired using the TV camera mounted on the carriage. This picture is analysed and the exact position of the DU is determined, enabling the robot to lower the gripper arm and grasp the DU. The gripper (Figs. 1 and 2), is fairly compliant having two thin fingers that slide round the ferrite yoke, and which can then be locked. As the gripper pulls the DU out of the box, a pneumatically operated 'pusher'

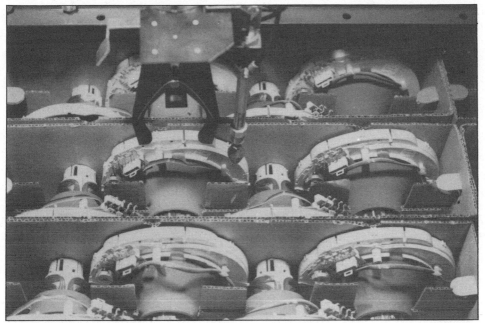

Fig. 1   Gripper over DU

holds the bottom of the box down. A built-in knuckle mechanism allows the DU to be placed horizontally onto the conveyor belt (Fig. 2). All the major functions of the gripper are checked by sensors.

Fig. 2   Gripper releasing DU

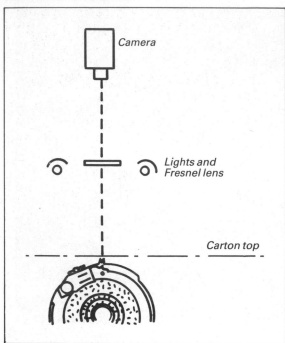

Fig. 3   Optical system for unpacking

Camera

Lights and Fresnel lens

Carton top

Location of the DU in the carton

The ease with which a picture can be processed is largely dependent on the degree of control that can be exercised on the appearance of the object and its background. If the object can be provided with readily visible reference marks of suitable geometry, or if the object can be silhouetted against a contrasting background, then simple methods can often be used. This is not the case with the DU unpacking task. The DU itself lacks specific reference marks, and the background is not ideal. The problem is complicated because the object distance varies with the layer in which the DU lies.

A solution is to equip the camera with parallel projection optics [4]. Apart from giving an image whose size is independent of object distance, other advantages of parallel projection include immunity to ambient illumination, ease of calibration, and elimination of parallax errors. Such optical systems have been used successfully in a number of applications [3,5], although mainly with back-lit objects. It was decided to use a similar technique, but relying on paraxial light scattered from the object (Fig. 3), as there is no possibility of using light transmitted from a source beneath the DUs. The resulting image is shown in Fig. 4.

A multi-pass picture-processing method is used, in which a number of features are identified in sequence. The method relies on the fact that the camera has been calibrated, as described below, and that the image size remains constant as the object distance varies. It is then possible to adjust the threshold to 'isolate' a feature of known size from its background, and to find its position. With this knowledge, a search can be made for the next feature, as both its approximate position and threshold are now known. This

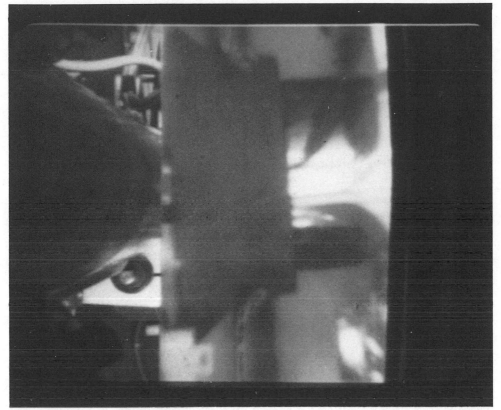

Fig. 4 TV camera image of DU

process can be repeated until the required feature is found. The value of this technique is that a progression can be made from features which, although reliably present and easily isolated, are not accurately positioned, to features which are difficult to find but which can be used to locate the object accurately. For the DU, the feature sequence consists of (Fig. 5):

- A highlight on the frame-coil winding.
- The edge of the ferrite yoke next to the winding.
- The edges of the coil against the background of the box.
- The edges of the yoke against the background of the box.

So the sequence starts with the location of an easily detected highlight, and ends with the accurate measurement of positions of visually obscure edges. The position and orientation of the DU can be calculated from the latter.

Performance

The system can remove a deflection unit from the box in approximately 10s. The required gripper position and orientation can be found in about 0.5s. The absolute errors found experimentally were less than 1mm in position, and 1° in orientation. These errors are acceptably small, especially considering the fact that the gripper is compliant.

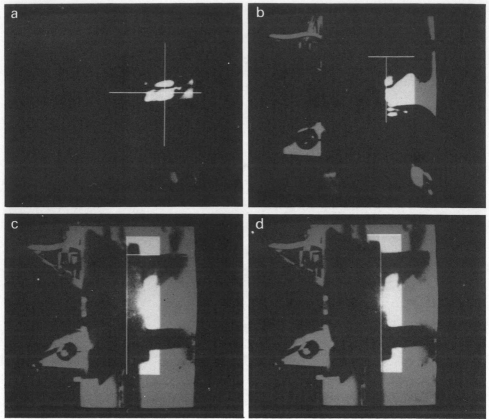

Fig. 5   Location of DU in the carton: (a) locating highlight, (b) locating yoke edge at centre, (c) locating upper edge of coil, and (d) locating upper edge of yoke

Mounting the DU onto the neck of the picture tube

The second part of the task is to place the DU onto the neck of the tube. It is necessary for this task to be performed quite independently from the unpacking, and the DU cannot be assumed to arrive in any particular orientation. The DU must be placed on the tube in a specific orientation, so that three 'landing areas' on the DU are exactly aligned with three ceramic studs on the tube. Once correctly placed, the DU must be pressed down onto the tube with a defined force while a screw clamp on the DU is tightened to a required torque, clamping the DU into position.

System structure and operation

The DU mounting task is accomplished by a PUMA 560 robot equipped with a specially designed gripper (Fig. 6) and three TV cameras to locate the DU and the neck of the tube. The gripper incorporates pneumatically operated clamping pads to grasp the top of the ferrite yoke, and a screwdriver to fasten the clamp. The required vertical, lateral and torsional compliance is provided by a combination of a compression spring module between the gripper and the PUMA wrist-flange, and 'metalastic' rubber mountings.

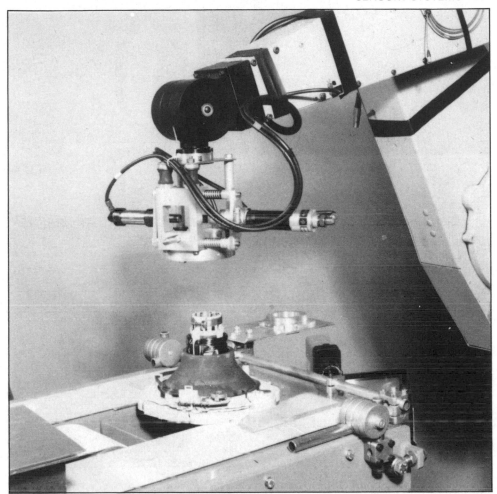

Fig. 6 PUMA about to pick-up DU

The unpacked DU arrives at the pick-up station on the conveyor belt. The vision system then uses an upward-facing TV camera mounted underneath the conveyor to determine the exact position and orientation of the DU, and the PUMA is instructed to pick it up. The PUMA orientates its gripper so that the screwdriver is aligned with the screw clamp, descends over the DU, grasps the ferrite yoke, and picks up the DU. Correct grasping is confirmed by a photodetector.

While the PUMA is picking up the DU, the vision system uses two cameras to locate the position of the neck of the tube. The DU is brought over the neck (Fig. 7), and lowered until the gripper's compliant mountings are slightly compressed. At this point, the DU is purposely in an 'under-rotated' orientation. A fixed rotation is then made that would overshoot a locking position between a recess on the DU and one of the studs. The engagement of stud and recess produces an impulse which is detected by an accelerometer to confirm proper engagement. The

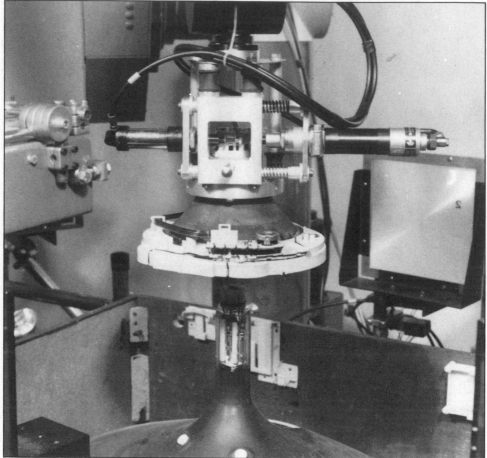

Fig. 7   PUMA about to place DU onto neck

over-rotation is accommodated by the compliant mountings. The screw clamp is then fastened to the required torque while the PUMA continues to hold down the DU.

Location of the DU at the pick-up station

The upward-facing camera is equipped with a parallel projection optical system, used in conjunction with both transmitted and reflected light to image all features of interest (Fig. 8(a)). Beneath the DU, and alongside the camera, two lamps illuminate the underside of the DU, while another lamp above it provides background illumination for the central hole in the DU. Parallel projection optics are used to avoid parallax errors which would otherwise occur if the axis of the DU were not aligned with the opitcal axis of the TV camera.

The picture processing uses the same feature-sequence technique used to locate the DU in the carton. The features used are the central hole, the edges of the copper winding where they lie on the plastic carrier, and three small holes in the plastic carrier.

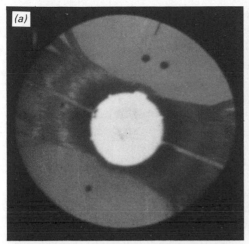

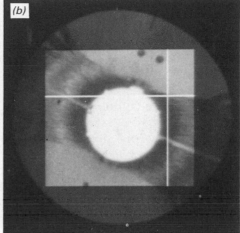

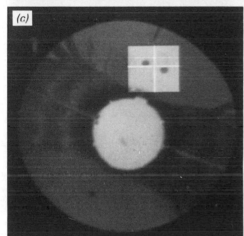

Fig. 8   Location of DU at the pick-up station: (a) underside of DU, (b) locating windings, and (c) testing for holes

The arrival of a DU is detected by making a scan to cover the central hole and neighbouring windings. A check on the range of light levels in the scan reveals the presence of an object. A threshold is selected which isolates the hole pixels against the background pixels. A check on the threshold level and the centre of gravity of the hole pixels indicates that the object is a DU. Further scans are made, and the centre of gravity is measured again. Coincidence of two successive measurements indicates that the candidate DU has stopped moving, and the centre of gravity is then stored as the DU position in the TV coordinate frame.

The orientation of the DU is found in two further steps. The first step is to examine a circular annulus of pixels centred on the centre of gravity of the hole. The annulus cuts through the winding and its background, and simple thresholding reveals the positions of the winding edges (Fig. 8(b)). After checking that the winding width is within an allowed tolerance, the edges are used to determine the orientation of the DU, but with a 180° ambiguity arising from the symmetry of the winding pattern. The second step is to position two small scans over the small holes in the plastic carrier (Fig. 8(c)). The 180° ambiguity can be resolved by determining which scan contains the single hole, and which scan contains the double hole.

Location of the tube neck

The location of an object in space from two views requires that the projection of a common object point be identified in each image. Both TV cameras used to image the neck of the tube are equipped with back-lit parallel projection optics. Each image (Fig. 9) is thresholded at a level which gives the correct neck width in the resulting binary representation. The centre of the top of the rectangle, which corresponds to the projection of the centre of the top of the neck is found in both images (Fig. 10). The intersection of the two rays passing through these points and parallel to the appropriate optic axes, gives the location of the top of the neck in space.

Performance

A DU can be mounted onto a picture tube in approximately 10s. Location of the DU and the tube neck each take about 1s, once the part is stationary. This is not critical, as picture processing takes place in parallel with robot movement.

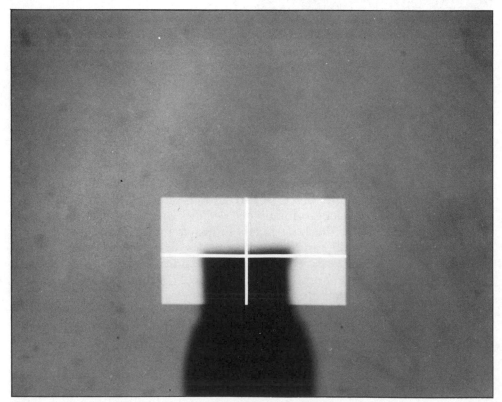

Fig. 9   Image of neck

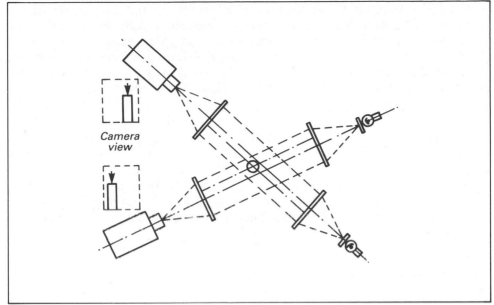

Fig. 10   Location of neck from two views

Calibration of TV cameras

Calibrating a TV camera equipped with conventional perspective projection optics usually means determining its position and orientation in 3D space[6]. Using parallel projection, it is sufficient to determine the camera orientation relative to 'world coordinates', and the magnification of the lens.

Fig. 11   Image of calibration rod

The TV camera in the unpacking machine moves in X, Y and Θ with the gripper. It is calibrated using a special plate, carried by the gripper, and which can be swung into the field of view of the TV camera by a pneumatic actuator. The plate is black with a pattern of four white spots, chosen for ease of picture processing. They allow the position of the gripper in the TV image to be found, and the X and Y spacings of the white spots permit the scaling factors to be calculated.

In the mounting system, each camera is calibrated using a pointed cylindrical rod which is carried by the PUMA. The spatial offset between gripper and rod-tip is known. The rod-tip is moved into the field of view of each of the three cameras in turn. The position of the rod-tip in the TV image (Fig.11) is determined using a simple recognition algorithm. The rod-tip is then moved through known distances in X, Y and Z, and is relocated using the same algorithm. The transformation matrix can then easily be calculated.

Control by ROBOS

Both the unpacking system and the mounting system are controlled by software based on ROBOS. ROBOS is conceived as a robot operating system by analogy with a computer operating system, in that it contains a number of facilities which are generally required in the development of sensor-controlled robot systems. ROBOS also provides a framework and discipline to guide the producer of the task-specific parts of the software, giving the whole activity greater coherence. It does not, however, contain computer operating system facilities such as handling files or running compilers.

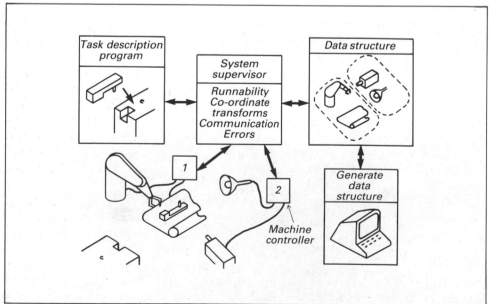

Fig. 12   Structure of ROBOS

At the core of ROBOS is a data structure which enables the user to model both the processing structure and the physical structure of the system being developed. The user can describe the processing structure in terms of 'machines', which are virtual processors, and the 'functions' which are to run on each machine. The physical structure can be described in terms of 'devices', which are collections of sensors and effectors, and 'frames of reference', in which positions and motions can be described.

For example, consider the two robot systems described above, each having TV cameras attached to them, but able to share the same vision processor. From a processing point of view the cameras are part of a single machine capable of executing vision functions. From a physical point of view they are separate devices, each with its own frame of reference.

The structure of ROBOS is shown in Fig. 12. At the centre is the system supervisor and the data structure. The supervisor is connected to the task description program and also to the specialised machines, each of which is controlled by a software module called a machine controller. Machines are connected by standard ROBOS communication channels, which are implemented by shifting data in memory, or by a physical connection if the machine controller is installed in a separate computer.

When the task description program requires a function to be performed it calls the supervisor, which first examines the data structure to determine the 'runnability' of the function, i.e. whether any relevant device is in such a state as to block execution of the function. Two options are available when a function is not runnable. Control can be returned immediately to the task description program, or the supervisor can wait for a specified time to see if the function becomes runnable, failing which it times out and creates an error condition. The supervisor performs all necessary transformations between frames of reference, as each machine controller expects to receive location coordinates in its own base frame rather than that specified in the function.

Machine controllers do not have a fixed structure. Some controllers may be programmed quite independently. For example a PUMA robot with a suitable VAL program is quite acceptable as a ROBOS machine controller. In general a machine controller must be able to accept commands with locations expressed in its own base frame, return results, and indicate when the execution of a function has been completed and whether it was completed successfully. The manipulator controllers which we have written contain a specific 'geometric model' for converting between the base frame and actuator values. This makes it simple to change from one robot geometry to another.

Once a function which involves motion of part of the system has been executed, the relationship between frames of reference may no longer be valid. ROBOS automatically updates these relationships whenever necessary.

The handling of errors or exceptions in flexible assembly systems often presents difficulties. For example, it is not easy to decide at which level error recovery procedures should operate. ROBOS distinguishes between fatal errors which halt the system and non-fatal errors from which recovery is possible. Recovery strategy tends to be task dependent, and therefore ROBOS contains a mechanism to return control to a specified part of the task description program in the event of a non-fatal error.

References

[1] Hale, J.A.G. and Saraga, P. 1975. Control of a pcb drilling machine by visual feedback. In, *4th IJCAI*, Tblisi, pp. 775-781.
[2] Saraga, P. and Skoyles, D.R. 1976. An experimental visually controlled pick and place machine for industry. In, *3rd IJCPR*, Coronado, pp. 17-21.
[3] Jones, B.M. and Saraga, P. 1981. The application of parallel projections to three-dimensional object location in industrial assembly. *Pattern Recognition*, 14 (1-6): 163-171.
[4] British Patent Application No. 2065299A, 1979. *An Object Measuring Arrangement*.
[5] Jones, B.M. and Saraga, P. 1980. Simple 3D assembly using parallel projection optics. In, *5th IJCPR*, Miami.
[6] Ballard, D.H. and Brown, C.M. 1982. *Computer Vision*. Prentice-Hall, Englewood Cliffs, NJ, USA.

VISION-GUIDED ASSEMBLY OF HIGH-POWER SEMICONDUCTOR DIODES

J. J. Hill, D. C. Burgess and A. Pugh
University of Hull, UK

First presented at the 14th International Symposium on Industrial Robots and 7th International Conference on Industrial Robot Technology, 2 – 4 October 1984, Gothenburg, Sweden. Reproduced by permission of the authors and IFS (Conferences) Ltd.

A multisensory workstation has been developed for carrying out the batch assembly of power diode units using a PUMA 560 robot. The sensory system includes three vision sensors. An overhead work monitor camera is used for coarse component location and checking assembly operations, a low-resolution gripper camera is used for fine component location and inspection, and a camera inspection station is used for inspecting the terminating 'C-crimp' and locating its end-point. The hardware and software features of the sensory robot system are described and assembly success rates are compared with a non-sensory assembly of the same devices.

The manufacture of high-power semiconductor devices, such as thryristors and power diodes, occurs in small batches and is currently a labour intensive area of industry where there is considerable potential for automating production using flexible manufacturing systems. Such systems require the use of robots and associated sensors to automate the assembly operations. Research on sensor-guided assembly has been in progress at the University of Hull since 1981 and part of this research has involved collaboration with Marconi Electronic Devices who have supplied a number of 'model' assembly problems to guide the direction of the research. Some of this work has been reported elsewhere[1,2] but this paper describes progress on one of the assembly problems, the power diode unit.

Problem description

The manufacture of power diodes (see Fig. 1) involves constructing an active subassembly, shown in Fig. 2, for the 70A specification diode. Each diode unit comprises five components which, together with a graphite cylinder or 'weight' must be assembled into a graphite 'mountdown' jig, with 10-15 units

Fig. 1   Diode units

per jig depending on the diode specification. The graphite jig and weights hold the subassemblies in place as they pass through an oven where the solder melts to bond the components together.

The active diode pellets are either circular or hexagonal in shape and are fabricated by etching from a silicon wafer. The pellets are often poorly formed, containing approximately 25% reject shapes, raising the problem of component inspection and presentation which cannot be solved by traditional 'hard' automation methods because of the batch size.

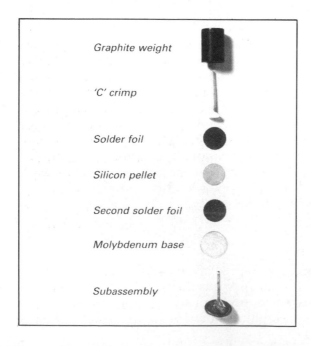

Fig. 2   Component parts for 70A diode unit

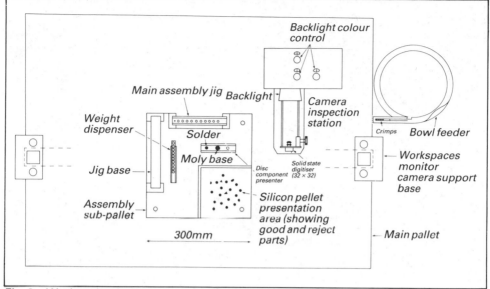

Fig. 3   Workstation layout

The assembly is automated using a PUMA 560 robot which has the accuracy required for this task and can easily be programmed using the VAL language[3]. The cycle commences with the assembly of weights into the holes in the assembly jig. The 'C-crimps' are then inserted into holes in the weights, followed by the solder foils, diode pellets, second solder foils, with the molybdenum bases assembled on top. Finally the jig base is fitted on top of the assembly jig and the completed assembly turned over before it is transported through the oven in the next stage of the manufacturing process.

The workstation

The work area, shown schematically in Fig. 3 and photographed in Fig. 4, consists of a small aluminium pallet holding all the fixed part presenters which locates into a larger fixed pallet by means of four dowels. This flexible approach allows the rapid changeover of the robot from one task (in this case, the manufacture of power diode units) to another (e.g. the thyristor assembly). The small pallet contains a gravity feed dispenser for the graphite weights, and magazines for the solder foils and molybdenum bases. Inspection of all these components is unnecessary and they are assembled 'blind' into the jig. The assembly jig and high base are held in cradles on the small pallet which also contains a fixed area for the presentation of the silicon diode pellets. The C-crimps are fed from a bowl feeder located off the main pallet.

The gripper

The gripper attached to the end of the PUMA arm has three functions: to pick up the silicon discs, the solder foils, the molybdenum bases and the C-crimps by means of a suction gripper; to pick up the weights by means of a

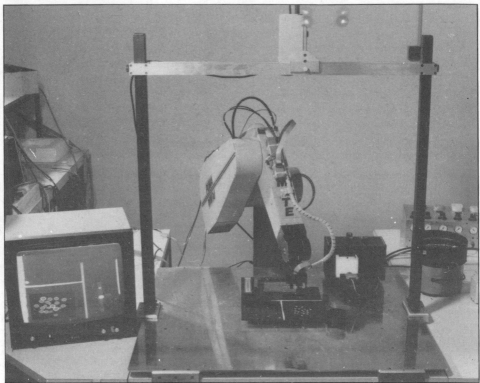

Fig. 4   General view of assembly workstation showing workspace monitor camera, gripper camera and camera inspection station

standard parallel jaw with an indentation to enable them to be held securely; and to pick up the jig base using a fixed jaw and one of the moving jaws.

A vacuum generator powered by compressed air operates the suction gripper, with the air turned on and off by a solenoid operated valve controlled from within the VAL program.

The gripper also carries a low-resolution camera, and this is discussed in the next sections.

Sensing requirements

It was established early in the project by attempting a 'blind' assembly of the devices, that a multisensing approach would be necessary in automating the assembly. The sensory system that has been adopted includes three vision sensors, a vacuum sensor and a proximity sensor. The role of each of these sensing mechanisms is now described.

The C-crimps are retrieved from the track at the end of the bowl feeder using the same vacuum 'sucker' as is used for picking up the disc components. There is a small amount of error as to the positioning of the crimp on the end of the sucker and in addition the shank of the C-crimp is sometimes bent or not perpendicular to the contact end. For these reasons, and because of small robot inaccuracies, a reliability of only 81.5% was achieved for a blind insertion of the C-crimp into the hole in the weight, even with a large chamfer on the hole. A camera inspection station (CIS) is located on the main pallet (see Figs. 3 and 4) and consists of a very low resolution (32 × 32) image digitiser together with a backlight. The CIS is used to inspect a C-crimp as it passes from the bowl feeder to the assembly jig and to determine the position of the tip of the post relative to the robot's frame of reference.

A vidicon camera is mounted directly above the workplace (see Fig. 4) and is referred to as the workspace monitor camera (WMC). This is a 256 × 256 pixel, grey-level, camera system that has the function of isolating and providing positional information on all objects within the silicon pellet presentation area. It can additionally be used for checking the fixed part presenters and for inspecting each stage of the assembly.

Another very low resolution (32 × 32) vision sensor called the gripper camera (GC) is mounted on the robot end-effector in a small 36 × 30 × 18mm housing which does not significantly affect the payload of the robot. Both these 32 × 32 pixel vision sensors are of the dynamic RAM type. The GC has the functions of providing a fine adjustment of position, and component inspection. It is able to do this because, despite its low pixel resolution, it provides an image with a higher spatial resolution than the WMC due to its close proximity to the part. The image obtained with the GC is very sensitive to fluctuations in light intensity (e.g. a shadow caused by the robot). In order to eliminate this a fibre-optic light source is incorporated into the camera body.

The non-vision sensors used are a crossfire sensor, consisting of a light-emitting diode-photodetector pair, which is used at the end of the bowl feeder track to indicate when a C-crimp is waiting in the pick-up area, and a vacuum sensor to detect if a part has been picked up.

System architecture

From the previous section it is evident that the robot system must be able to cope with a number of sensory inputs. The advantages of the multi-processor approach to multi-sensing, where each sensor and/or actuator has its own

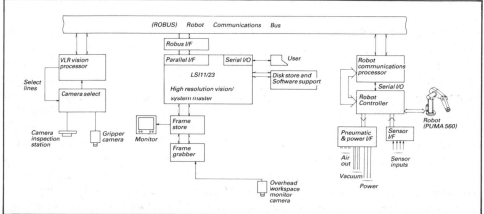

Fig. 5   Multisensor system architecture

local data processing unit, has been previously reported[4,5]. Accordingly, it was decided that this type of approach would be adopted for the diode assembly with data processing software and special I/O requirements for given sensors consigned to an array of dedicated slave processors. A master processor is responsible for communication between the user, robot and slave subsystems with the slave processors acting as intelligent interfaces between sensory input devices, workstation actuators and the robot controller. Sensory tasks are initiated by the master processor and are executed by the appropriate slave sub-system as and when required within assembly operations. Inter-communication is carried out by means of 'semaphore' messages passed via a parallel bus architecture, ROBUS[4]. A task 'tree' which optimises the use of concurrent processes, is loaded into the master processor which then performs the assembly operations by stepping through the task tree and initiating sub-tasks as necessary.

The specific system architecture used for the power diode assembly is shown in Fig. 5 in which three processing subsystems are interfaced to ROBUS:

- The low-resolution vision processor (Z80).
- The system master/high-resolution vision processor (LSI11/23).
- The robot communications processor (8085).

The low-resolution vision processor can handle up to four camera inputs and in this application is used to process 32×32 pixel images from either the CIS or the GC. The image to be analysed is selected under software control.

The LSI 11/23 carries out the high-resolution vision processing and also acts as the system master. The microcomputer is fitted with $256 \times 256 \times 8$ bits framestore/frame capture boards for high-resolution vision processing, RLO2 disks for the software support and a DRV11 to ROBUS interface. Multitasking software is used to carry out the two functions of system master and high-resolution vision processing on the same processor.

The robot communication process (RCP) is a slave system used to enable efficient communication with the robot. The developments described in this paper were initiated before the introduction of VAL2, with the main disadvantage that the provision for sensory feedback to the PUMA robot was very limited. The robot communicates with the RCP by both parallel and serial I/O channels which are interfaced within the robot controller to the VAL operating system. The parallel I/O allows for 8TTL inputs and 8TTL outputs to be linked between the two systems and the serial channel is an RS232 compatible bidirectional link running at 9600baud. Simple flag information is passed between the robot controller and the multiprocessor via the TTL lines, while more complex information is sent via the serial link directly into the VAL command console input channel. The serial channel is configured as a terminal emulator and the command strings VAL receives are interpreted as originating from a user terminal. By employing this technique it is possible to overcome some of the problems which arise in attempting to pass symbolic information into a VAL program via the more usual I/O module interface and it is possible to redefine locations or modify actions during program execution.

The task of communicating with the PUMA was consigned to a slave processor because of the way the PUMA controller deals with the serial interface and, because of the complexity of analysing the character stream coming back from the VAL software. Although the serial channel is specified as 9600baud this only refers to the character frame rate; the inter-character transmission time is indeterminate, being a function of the amount of processing required to analyse the previous characters in the current string. It is for this reason that the RCP is required to handle all the necessary string buffering, determine current robot status, and perform error message analysis whilst keeping these processes transparent to the system master processor. Flag and serial string information is transmitted to and from the robot on demand from the master processor, enabling the robot to be directed as the result of decisions taken by the master processor based on sensory input from other slave-processors within the system. Work carried out at Hull to modify the PUMA software, (VALAD), gives partial access to the parallel I/O interface for the passing of parameters into VAL. By this technique it is possible to increase the speed at which the robot may be directed by the external sensory input, allowing for rapid sensory based control to be employed.

Vision algorithms

C-crimp end detection

An image is taken of the backlit C-crimp shank. This image is then scanned from top left to bottom right. For every line that contains just a single run of data (i.e. background, object, background), the mid-point of the run is calculated and stored. If the line contains no object, or more than one run of data, it is assumed to have an invalid mid-point. This list is then scanned back to find the first valid reading, which will be the end point of the crimp. Note, that this data also gives a measurement of the angle which the C-crimp makes the vertical. The difference between the end-point and the centre of the image is then transmitted to the robot as an error signal, and the process repeated until the error is zero. The C-crimp is then rotated through 90° by the robot and the process repeated.

Isolation and position of parts in presentation area

This breaks down into two stages, firstly the separation of the grey-level picture into a binary one, and secondly the analysis of the binary image to produce meaningful object data. The vision processing is summarised in Fig. 6.

The processing window is previously defined in an earlier set-up phase. A histogram is made of the pixels within this window which is then smoothed by a two-pass algorithm and examined for peaks. If more than two peaks remain, the process is repeated. The mid-point of the two maxima is then found and returned as the threshold value. The area of the image within the window is then thresholded at this value.

In the second stage image analysis, the image is scanned left to right, top to bottom within the window until a blob is found. First a check is done to make sure that this object has not been detected before. If not, then the perimeter

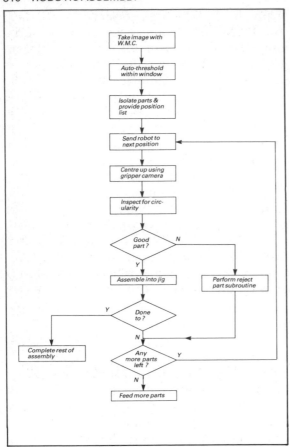

of the blob is traced out, at the same time marking the points on screen. If the object has already been marked, then the scan is continued until another blob is found, or the end of the window is reached. For each object found the following information is evaluated:

- Initial starting point on the perimeter.
- The chain code.
- Perimeter value.
- Area (in pixels).
- x length of bounding box.
- y length of bounding box.
- y position of centroid.
- x position of centroid.
- Minimum y coordinate of bounding box.
- Minimum x coordinate of bounding box.
- Maximum y coordinate of bounding box.
- Maximum x coordinate of bounding box.

The resulting screen display is shown in Fig. 7. The centroid coordinates of objects which exceed a minimum area are then passed to the robot.

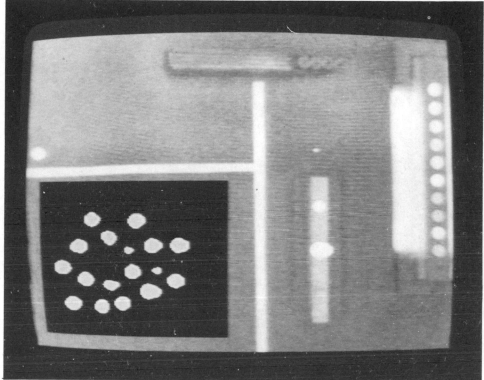

Fig. 7 Workstation monitor display after processing to locate silicon pellets within an image window

Gripper camera fine tuning and silicon pellet inspection

The robot moves to a position in the work area such that the gripper camera is above the object of interest (see Fig. 8). An image is taken from this camera and the centroid calculated. The difference between this and the centre of the image is found, and transmitted to the robot as an error signal. Another image is then taken and the process repeated until the centroid corresponds to the centre of the image.

Once the GC has been exactly centred above the silicon pellet the inspection task takes place. This is done in the following way:

- Take an image.
- Rotate it in software by 90° about the centre of the image.
- Compare the rotated image with the original image, placing the result in a third image area.
- Remove all isolated and diagonally connected pixels from this image.

For a perfect circle the result should be a blank image. In practice due to quantisation errors etc. after stage 3, a ring of pixels remains evenly distributed around the perimeter which is easily removed by stage 4. For a reject part, the areas remaining after stage 3 are much more connected than for a good part, and areas remain after stage 4 is executed.

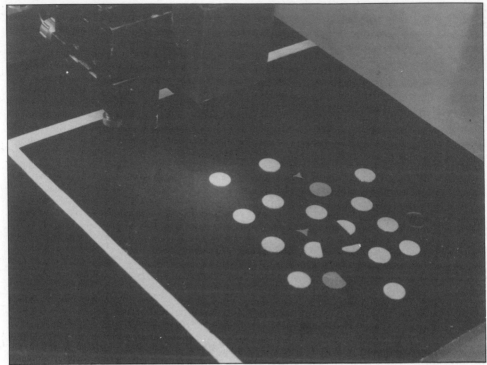

Fig. 8   Silicon pellet location and inspection by gripper camera

Software structure and multiprocessor multitask control

All the following functions must be bound together with a suitable software structure:

- Communication with the robot.
- High-resolution vision processing.
- Low-resolution vision processing.
- Overall coordination (master).

For the purpose of development in a research environment, an interpretive command language was developed which allows ideas to be evaluated and modified very easily. Multiprocessor multitask control (MCP)[6] allows the overall coordination of all sensor/actuator functions, be they implemented on separate processors interfaced to the ROBUS, or directly interfaced into the master processor. It runs on the DEC 11/23 under RT11, and uses floppy disks for program and data storage. It has the following facilities:

- Communication with a user console.
- Direct communication by the user with any slave processor on ROBUS.
- Direct communication by the user to a second serial line.
- Entry and execution of organisational programs.
- Task initiation, parameter passing, protocol, etc.
- Simple arithmetic.

```
;+
;                   This program centres the GC above the silicon pellet.
;-

        LET SCA=12                  ; Set a scaling factor.
        LVIS 11,1                   ; Select correct camera.
RPT:    LVIS 2                      ; Take a picture.
        LVIS 10/XCE,YCE,ARE         ; Find X and Y centres, and area.
        LRSTAT                      ; Wait for routine to complete.
        IF ARE>0; MOV               ; If area non-zero, then move robot.
        PRINT "No object";<13>;     ; Try again.
        GOTO RPT
MOV:    PRINT XCE,YCE,ARE           ; Display image received.
        LVIS 8,1
        LET XCE=XCE-16*SCA          ; Set up for robot movement.
        LET YCE=YCE-16*SCA          ; by adjusting XCE, YCE.
        LRSTAT                      ; Wait for display task to finish.
        IF XCE <>0; MV2             ; X centre non-zero, so move.
        IF YCE <>0; MV2             ; Y centre non-zero, so move.
        PRINT "Centred";<13>; END   ; Exit program.
MV2:    RCOM 63,XCE,-YCE            ; Now move the robot.
        RCSTAT                      ; Wait for motion to finish.
        GOTO RPT                    ; Check again.
```

Fig. 9 Sample program segment

The names for variables and labels can be up to three characters long and any valid RAD50 character can be used except space. Variables are integers in the range −32767 to +32767 (the value −32768 being used to identify when a variable has been declared but not assigned a value). Labels are terminated by a ':' followed by a space. A command need only have sufficient characters to make it non-ambiguous. A sample segment of program is given in Fig. 9.

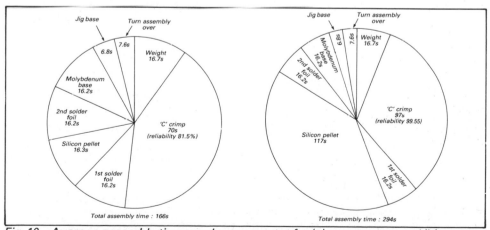

Fig. 10 Average assembly times and success rates for (a) non-sensory and (b) sensory assembly of 10 diode units

Concluding remarks

At an early stage in the research the diode assembly was attempted 'blind' so that by taking reliability measurements it was possible to identify areas where 'sense' could improve assembly success rates. Fig. 10 summarises assembly times and reliabilities for the non-sensory and sensory assemblies, respectively.

For the non-sensory parts of the assembly, that is, the assembly of the weight into the jig, assembly of the solder foils, molybdenum base and jig base, better than 99.5% reliability can be achieved. This is due to rigid presentation and shaking of the jig to ensure that parts settle correctly. For the non-sensory assembly of the C-crimp a reliability of 81.5% was the best achieved. By introducing visual feedback this has been increased to better than 99.5% at the expense of a 38% increase in time for this part of the assembly. The non-sensory assembly of the silicon pellets was carried out by manually selecting perfect components and presenting these in a stack adjacent to the molybdenum bases. The automated assembly of the silicon pellets is not possible without vision or some other form of sense. The increase in time from 16.3 to 117 is mainly caused by increased time in moving and in communicating with the robot, rather than in vision processing. Although the reliability of the sensory diode assembly is good, the overall assembly time of 294 (for 10 devices) would not be acceptable in a production environment. However, work is in progress in speeding up parts of the assembly which, coupled with a faster robot, should make the approach economic.

Although the multisensory workstation described in this paper has been developed for the assembly of power diode units, with minor modifications the sensory system can be used for assembling other high-power semiconductor devices and research is also in progress on investigating the generic applications of this and other sensory work.

Acknowledgements

The authors acknowledge, with gratitude, the support of the Science and Engineering Research Council under grants GR/B/18693. Useful discussions have also taken place with their industrial partners for this project: GEC Research Laboratories and MEDL.

References

[1] Burgess D. C., Hill J. J. and Pugh A. 1982. Vision processing for robot inspection and assembly. In, *SPIE Proc. Robotics and Industrial Inspection*, Vol. 360, pp.272-279.

[2] Taylor P. M., Selke K. K. W. and Pugh A. 1983. Visual feedback applied to the assembly of thyristors. *Congress AFCET Autamatique*. Productique Intelligente, Besancon, France.

[3] *Users Guide to VAL, a Robot Programming and Control System*, Version 12. Unimation Inc., Danbury, CT, USA, 1980.

[4] Mitchell I., Whitehead D. G. and Pugh A. 1983. A multiprocessor system for sensory robotic assembly. *Sensor Review*, 3(2): 94-96.

[5] Whitehead D. G., Hill J. J., Mitchell I. and Burgess D. C. 1984. Industrial robotic assembly: A multi-sensor approach. In, *IEE Int. Conf. on Flexible Automation Systems*. IEE, London.

[6] Burgess D. C. 1984. *Multiprocessor, multitask control (MPC): Documentation*, Internal Report. Robotics Research Unit, University of Hull, UK.

6

Economics

Complementing the more general approach of previous sections, this chapter presents three different approaches in the consideration of robotic assembly economics. The contributors consider comparative costs, and then examine the financial case for general purpose and multi-arm assembly robot systems.

ROBOTIC ASSEMBLY SYSTEM DESIGN, ASSEMBLY COST, AND MANUFACTURING VIABILITY

J. Miller
Hatfield Polytechnic, UK

First presented at UK Robotics Research 1984, 4-5 December 1984, London. Reproduced by permission of the authors and The Institution of Mechanical Engineers.

A practical and realisable robotic assembly system based on Fanuc assembly robots, a Bosch FMS conveyor, and vibratory-bowl parts feeders is discussed. The system is conceived as an in-line arrangement of cells, the cells being linked by the conveyor. Each cell contains one robot, and a number of parts feeders. The merits of 'simple', 'universal' and 'turret' grippers are examined. Whether finished assemblies should be manually or automatically unloaded is evaluated, as are manpower requirements. The shift working and capital recovery period necessary to make the system competitive with manual assembly are discussed. The aim of the work is to define how an in-line system should be designed and run. It is concluded that, at current costs, two or three shift operation is essential, and that an output of several thousand assemblies per day is needed to justify the required capital investment. It is shown that economic viability is very sensitive to product design. Turret grippers offer advantages, but there is little to choose between manual and automatic unloading.

Recent progress in automated piece-part manufacturing has not been matched by as widespread automation of assembly. The reasons for this disparity are mainly economic, though as moderately priced robots proliferate, the prospects for automated assembly might be thought to be improving. Automated assembly has a formidable competitor in manual assembly, however, which requires little capital investment, and whose flexibility and dexterity are hard to emulate in a machine. For the general run of production, automated assembly will be judged primarily against cost criteria, since manual assembly is able to overcome most technical difficulties.

It is reasonable to assume that robots can be made to do a range of simple assembly tasks, if only because special-purpose automated assembly machines can accomplish a similar task with far less manipulation ability. What is less obvious is the outcome when robots work in conjunction with other necessary equipment in an assembly system. The aim of this paper is to examine how a robotic assembly system should be designed and operated so as to achieve economic automatic assembly, using currently available system components.

Accounting aspects of substitute capital equipment for manual labour

Expenditure on manual labour and expenditure on capital equipment are different in nature. Labour cost is a series of small payments, usually weekly, spread uniformly over the period of operations. Capital cost can be taken as a single large payment at the start of operations. Both, of course, must be recovered from revenue arising from sales of the product produced.

When, as with assembly processes, manual and machine operations are to a large degree interchangeable, means must be found of setting the two kinds of expenditure on the same accounting basis. Boothroyd's approach[1], which is followed here, is to define a 'breakeven capital cost', Q. A machine of cost Q is identical in cost terms to a manual operator employed for one shift per day.

The procedure adopted here to calculate Q is to equate to zero the net present value of the sum of capital cost and suitably discounted expected savings and recoveries. In calculating the values given in Table 1, the following specific assumptions are made:

- Operator wages are paid weekly and discounted in weekly periods.
- Wage inflation occurs at an annual rate of 5% or 15%.
- Future savings or recoveries are discounted at an annual rate of 10% or 15%.
- The book value of capital equipment is written down linearly to nil over 4 or 8 years.
- Recovery from resale or reuse of capital equipment is at 50% of book value in the year of disposal.

Table 1   Breakeven capital cost Q as a multiplier of annual employment cost of the weekly paid manual assembly operators

		Case A	Case B	Case C	Case D
Discount rate (%)		10	10	15	15
Annual wage inflation (%)		5	5	15	15
Capital writing down period (years)		8	4	8	4
Full years of operation of equipment			*Multiplier Q*		
	1	1.58	1.44	1.50	1.38
	2	2.69	2.34	2.58	2.27
	3	3.55	2.99	3.47	3.00
	4	4.25	3.53	4.26	3.65

- Investment in capital equipment is funded from reserves and not by borrowing.
- Recoveries of corporation tax allowances on capital equipment are ignored in line with future provisions in the UK 1984 fiscal budget.

At present, most companies expect capital investment projects to be self-financing within a two-year period, even when the working life of the equipment is much longer. On a two year basis, it would therefore be justified to invest between 2.27 and 2.69 times the annual cost of a manual operator to reduce manpower by one operator per shift. Longer capital recovery periods are obviously desirable where they can be justified.

Assembly system layout

The assembly system layout adopted for analysis is shown in Fig. 1. It is an in-line arrangement of a series of workstation cells. Each cell contains one assembly robot, together with a number of parts feeders. The workstation cells are connected by a conveyor system in rectangular layout. Part-completed product assemblies are transported through successive worksta-tions on workcarriers.

When product assembly is completed at the last workstation, the finished assemblies must be unloaded from the assembly system. Two alternatives for unloading are envisaged. In the first, the robot in the last workstation automatically unloads assemblies from the workcarriers before they return

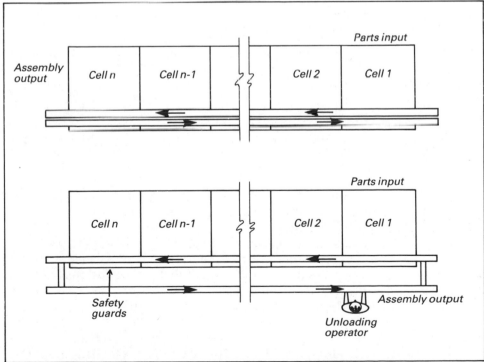

Fig. 1 Robotic assembly line layout

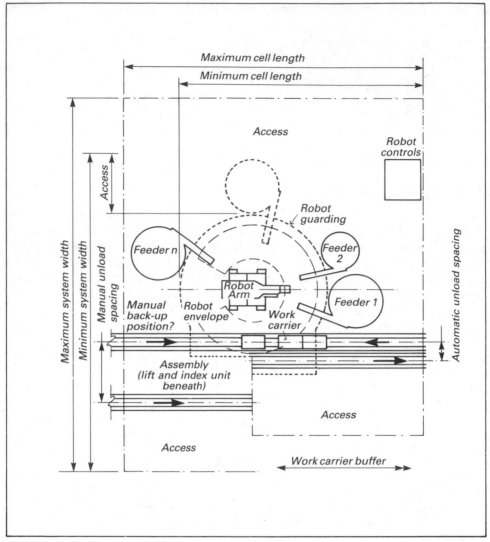

Fig. 2   Assembly cell layout

to the first workstation. In the second alternative, the finished assemblies pass to an unloading station in the return leg of the conveyor, where they are unloaded manually by an operator.

All workstation cells are nominally identical, as it is wise to avoid too much diversity of equipment types, to minimise spares inventory and maintenance costs on expensive capital equipment. The internal arrangement of a workstation cell is shown in Fig. 2. Parts feeding by vibratory bowl feeders is assumed, with the feeders arranged around the robot, in the sector of the robot's working envelope not occupied by the conveyor. It is essential to provide man access between the workstation cells and around the system, for both operation and maintenance. Easy access is particularly important during operation for clearing jammed parts, and to replenish feeders.

The in-line layout is adaptable to a wide variety of assembly sizes and rates of output, by suitable choice of the number of workstations and the number of parts assembled by each workstation. Generalised analysis is also easy, since all such systems are similar, whatever their size and output.

Basis of analysis

The in-line arrangement of robotic workstations is similar to the multi-station hybrid system analysed by Boothroyd et al.[1]. A 'hybrid' system is one in which small indexing machines are linked by buffered free-transfer conveyors. The robot in a robotic workstation performs the dual function of parts insertion and linking of feeders to the assembly in a fixed cycle, in a manner analogous to an indexing machine, and can be analysed on a similar basis.

The calculation of system economic performance is carried out using a specially written computer program[2]. An example of program output is shown in Fig. 3. By specifying in turn a variety of system configurations and computing the resulting assembly cost, the system designs which reduce assembly cost can be identified.

Assembly robots

A wide range of assembly robots are now available, with various arm designs. This analysis will be based on cylindrical geometry robots, though the trends would apply equally well to SCARA-type robots.

Specific data is presented only for the Fanuc range of assembly robots (Table 2). Axes Z, T and R control gross end-effector position within the operating envelope. Axes A and B (when present) are wrist axes. Axes A and B in Models A-0 and A-1, and axes Z and A in Models A-3-0 and A-4-0, can be omitted, with up to 30% reduction in cost. Fig. 4 shows the arm designs.

Table 2   Fanuc assembly robots – maximum axes configuration

			Model		
		A-0	A-1	A-2	A-3
Cost (f) mid-1984		24,500	30,500	15,000	13,000
Maximum number of servo axes		5	5	4	4
Maximum axes configuration*		PRPRR	PRPRR	RPPR	RPPR
Positioning repeatability (±mm)		0.05	0.05	0.03	0.03
Payload (kg)		3	9	6	2
Outer radius of operating envelope (mm)		650	1070	638	563
Maximum travel in axis	Z (mm)	300	500	120	100
	T (deg)	300	300	290	290
	R (mm)	300	500	300	250
	A (deg)	300	300	300	300
	B (deg)	190	190	–	–
Minimum time (s) for full span	Z	0.5	0.83	0.3	0.25
movement in axis	T	2.5	2.5	1.93	1.93
	R	0.25	0.42	0.38	0.25
	A	3.33	3.33	1.0	1.0
	B	3.17	3.17	–	–

*R = revolute, P = prismatic. Read left to right from base to end of arm

```
TOTAL NUMBER OF PARTS IN ASSEMBLY                        :  15
DEFECTIVE PARTS CAUSING STOPPAGES                 (%)  :  1
LARGEST PART SIZE                                 (MM)  :  50
HEAVIEST PART WEIGHT                              (GM)  :  500
MAXIMUM INSERTION TIME                            (SEC)  :  2
TOTAL COST OF PARTS FEEDERS                        (£)  :  24375

UNLOADING METHOD                                         :  AUTOMATIC, AT LAST WORKSTATION

ROBOT TYPE                                               :  FANUC A-MODEL 3
GRIPPER TYPE                                             :  UNIVERSAL, EACH COSTING £3000
TOTAL GRIPPER TOOLING COST                        (£)  :  1600
SIMPLE GRIPPER COST, WHERE USED                   (£)  :  500

CONVEYOR TYPE                                            :  BOSCH FMS
WORKCARRIER DETAILS:   LENGTH                      (MM)  :  240
                      WIDTH                       (MM)  :  240
                      CODE MEMORIES                     :  1
                      TOOLING COST                (£)  :  150

OPERATOR ANNUAL COST:  1ST SHIFT                  (£)  :  8000
                       OTHER SHIFTS               (£)  :  10000

SHIFTS (8 HOURS EACH) PER DAY                            :  2
OVERTIME PER DAY                                  (HOURS)  :  0
SYSTEM OPERATING TIME PER DAY                     (HOURS)  :  15.5
BREAKEVEN CAPITAL COST                            (£)  :  19760
```

PERMISSIBLE SYSTEM CONFIGURATIONS AND COSTS

A = CYCLE TIME (SEC)	G = OPERATOR LOADING (%)	M = TOTAL CAPITAL COST (£K)
B = NUMBER OF WORKSTATIONS	H = TOTAL SYSTEM LENGTH (M)	N = SPACE COST (£K/YEAR)
C = PARTS ASS'D PER WKSTN	I = TOTAL SYSTEM WIDTH (M)	O = MAINTENANCE COST (£K/YEAR)
D = PARTS ASS'D-LAST WKSTN	J = NUMBER OF WORKCARRIERS	P = ASSEMBLY COST (PENCE)
E = MAX OUTPUT (ASMS/DAY)	K = ROBOT+GRIPPER COST (£K)	Q = ASS'Y COST/PART (PENCE)
F = NUMBER OF OPERATORS	L = TOTAL CONVEYOR COST (£K)	

A	B	C	D	E	F	G	H	I	J	K	L	M	N	O	P	Q
6.4	16	1	0	8338	3	81	48.9	4.1	133	248$	69	341	14	34	12.8	.85
13.8	8	2	1	3882	2	56	25.9	4.7	41	146	32	202	9	20	16.8	1.12
20.3	6	3	0	2636	1	76	20.2	4.7	23	107*	24	155	7	16	17.5	1.16
26.9	4	4	3	1989	1	58	14.4	4.7	14	74	18	116	5	12	18.2	1.21
33.6	4	5	0	1588	1	46	16.8	4.7	10	71*	17	113	6	11	22.6	1.5
40.6	3	6	3	1315	1	38	13.3	4.7	7	56	15	95	4	9	23.7	1.58

NOTES: 1. LAST CONFIGURATION FILLS SPACE AVAILABLE FOR PARTS FEEDERS
2. $ IN COLUMN K INDICATES SIMPLE GRIPPERS USED AT ALL WORKSTATIONS
3. * IN COLUMN K INDICATES SIMPLE GRIPPER USED AT ONE WORKSTATION

Fig. 3   Example program output

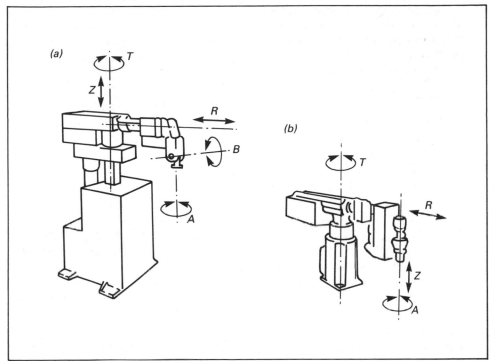

*Fig. 4   Fanuc assembly robot arm designs: (a) A-Model 0 and A-Model 1 and (b)
A-Model 3-0 and A-Model 4-0*

Ideally automatic assembly should be performed layer-fashion by inserting parts vertically from above. This restriction is well accepted for pick-and-place machines on dedicated automatic assembly systems, but often requires some product redesign. Pick-and-place machines normally have two or three axes. The one of two extra wrist axes available on robots permit part reorientation before insertion. Extensive reorientation may increase cycle times, however, if the slow wrist movements cannot occur in parallel with the gross movements.

When estimating cycle times advantage can be taken of the behaviour of the robot's point-to-point positioning system. If a movement to a new position involves displacements in two or more axes, the time required is determined by the speed of the slowest axis. For gross positioning movements, the R axis is much faster than the T axis. Thus small changes in R can mostly be neglected in estimating cycle times for assembly sequences. This simplification is used here in obtaining cycle times.

Parts grippers

When tooling the assembly system robots to handle individual parts, four strategies are available to cope with the diversity of part shapes:

- A 'simple' gripper, fixed to the robot, and able to grasp only few parts.
- A 'universal' gripper, which has jaws able to grasp every part encountered.

- A 'turret' gripper, which has multiple separately actuated sets of jaws mounted on an indexing turret.
- A 'hand-changing' gripper, where the actuator and jaws can be exchanged when needed. Such exchangeable actuator and jaw combinations are unlikely to be other than simple grippers.

With the last alternative, the hand-changing gripper, a time about the same as that needed to assemble a part is required to exchange the actuator and jaws. This time overhead, incurred at each exchange, must be added to the assembly cycle time, which, as will become evident, is uncomfortably long for robotic systems. As well as adding to cycle time, the hand-changing system increases capital cost and reduces payload. Taking all this into account, attention will be concentrated only on simple, universal and turret grippers.

The simple gripper is the lowest cost alternative, being usually an uncomplicated pneumatic device. Where a workstation handles only one part, there is little point using anything else.

The universal gripper must handle a range of part sizes, and so wide-opening parallel jaws are likely to be needed, with stepping motor or servo motor actuation. Design of the grasping surfaces of the jaws is complicated. An excellent account of the design considerations is given elsewhere[3]. As jaw separation varies from part to part the time to grasp or release a part will also vary.

A turret gripper must handle a similar range of parts. Jaw design is easier as each set of jaws is for one part only. Simple, pneumatically opened spring-closed scissor-type actuators can be used. Jaw opening is the minimum needed to clear the part, so grasp and release times are minimised. Extra time, however, is needed for indexing, which can be pneumatically or elctrically actuated, with a shot-bolt to preserve index accuracy. Turret grippers[4] are expensive and reduce payload, but can offer significant improvement in cycle time.

Workstation operating cycles and cycle times

Workstation cycle time is the sum of the times needed to acquire, transport and insert the parts assembled at that workstation. In an assembly system with in-line workstations, each workstation will, in general, assemble a different subset of the total parts comprising the complete assembly. The line must be balanced so that all workstations have approximately the same cycle time. The faster workstations must then be adjusted so that their cycle times are the same as that of the slowest workstation, to avoid the slowest workstation being a bottleneck to the flow of workcarriers through the system. Output is thus determined by the cycle time of the slowest workstation.

Here it is assumed that line balancing is approximated sufficiently well by making each workstation assemble the same number of parts. This requirement can be satisfied precisely if the total number of parts to be handled is an integer multiple of the number of workstations: if not, one workstation handles fewer parts than the others. For automatic unloading at the last workstation, the finished assembly itself is taken as a part to be handled by the last robot.

The cycle time of the slowest workstation is determined by the number of parts assembled there, and by the robot and gripper combination chosen for the system. The same robot is used at each workstation. That is also the case with grippers, except if any workstation handles only one part, when a simple gripper is used, so as to minimise total gripper cost. The assembly sequence differs with each type of gripper, and affects workstation cycle time in conjunction with robot speed.

The sequence of motions used to assemble a part with a simple gripper, assuming layer-fashion assembly build, is taken as:

- Part grasping, followed by slow Z-axis withdrawal from the feeder.
- Fast Z-axis movement to clear the feeder.
- Fast T-axis rotation toward the assembly.
- Fast Z-axis approach to the assembly.
- Slow Z-axis insertion into the assembly, followed by part release.
- Fast Z-axis movement to clear the assembly.
- Fast T-axis to return to the feeder.
- Fast Z-axis approach to the feeder.

The feeder should be positioned next to the conveyor track adjacent to the workcarrier buffer, to minimise the time of T-axis motion.

The sequence for the universal gripper is two or more repititions of the simple gripper sequence, since the universal gripper can transport only one part at a time. The order in which the two or more feeders are visited has no effect on total cycle time.

The turret gripper is able to transport as many parts as there are turret positions, because part grasping is by spring action. Here the number of turret positions is taken to be the number of parts assembled at the workstation. The turret can thus make a 'round trip', visiting each feeder in turn, and only then transporting the parts to the assembly. The fast and slow Z-axis motions are the same as for the universal gripper, but there is no 'to-and-fro' T-axis motion. Feeders must be arranged in the order in which parts are assembled, to obtain maximum time saving, but this is always possible. The feeder nearest to the assembly position can be visited first or last with no effect on workstation cycle time.

At the slowest workstation, fast motions in any axis are taken to be at the highest speed available from the robot. Workstation cycle time is thereby minimised. Times to withdraw parts from the feeder, or insert them into the assembly, will vary with part and assembly design; this is examined later. In order to estimate assembly system performance with the minimum of specific part data, all assembly sequence motions are taken to be uncorrelated random variates, with particular values chosen with uniform probability in prescribed ranged. This procedure models the variety of part shapes and ways of mating. Workstation cycle time is the sum of times for the several sequence components; its variance is much smaller than the variances of the individual contributions.

Parts feeding and orientating

It is assumed here that the continuous supply of orientated parts necessary for automatic system operation is provided by vibratory bowl feeders. Such feeders are available from a number of manufacturers. Bowl diameters of about 150, 300, 400, 500, 750 and 850mm are standard sizes. For successful feeding, bowl diameter should be about 10 times the largest dimension of the part to be fed. Bowl feeders are suitable, therefore, for parts with largest dimension less than about 85mm. In order to orientate parts, bowls are fitted with mechanical tooling specific to the part to be fed. Design and manufacture of orientating tooling is a highly skilled task. On average, about 40 technician hours are needed to tool a bowl for a particular part; difficult parts may need 120 hours. It is fortuitous if a bowl tooled for one part can be used to feed a different part.

Bowl feeder cost is the sum of the cost of the untooled bowl, vibrator and controls, and the cost of tooling. The first contribution depends on bowl size, and therefore on part size; the second is independent of part size. Typical total costs range from about £1450 to about £3300 for bowls between 150 and 850mm diameter, though prices for each size can vary widely.

An assembly system needs as many feeders as there are parts in the complete assembly, if all the parts are different. These feeders will be of different sizes, since most assemblies consist of a few large parts, some medium-sized parts, and a relatively large proportion of small parts. The total cost of all the feeders is needed for economic analysis, rather than an individual feeder cost. Total feeder cost is the product of average feeder cost and number of parts to be fed.

In order to estimate average feeder cost with the minimum of specific part data, part size distribution in a number of actual assemblies was investigated. Fig. 5(a) shows the normalised data; Fig. 5(b) shows the result of applying the empirical part size distribution of Fig. 5(a) to calculate average feeder

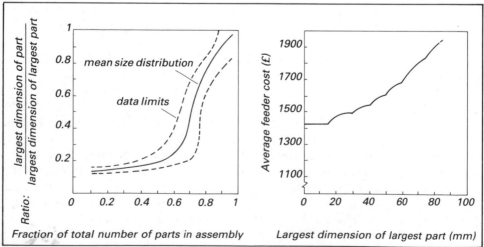

Fig. 5   Derivation of average bowl feeder cost: (a) normalised size distribution and (b) average feeder cost

cost as a function of the largest dimension of the largest part in the assembly. Despite its empirical nature, this procedure is unlikely to give rise to serious errors in estimates of overall assembly cost, since feeder costs rarely exceed one-third of total system capital cost.

The dedicated bowl feeder is a good average choice (see page 347). In some instances, however, other methods of parts presentation will be needed, for large, or fragile or difficult-to-orientate parts. Use of a few special feeders with a majority of bowl feeders should not affect overall assembly cost to any marked extent. Only genuine flexible feeders[5] able to handle a variety of parts, are likely to offer lower feeder costs; such feeders are not yet available to the general user.

Conveyor system

The function of the conveyor system is to transport partially completed assemblies from one workstation to the next. The conveyor can be an indexing or a buffered free-transfer conveyor. If workstations are susceptible to stoppages (e.g. due to defective parts) and workstation cycle time is small, the buffered free-transfer conveyor can provide much higher system output than an indexing conveyor, especially for assemblies containing a large number of parts. Buffered free-transfer conveyors are preferable for this reason, as well as for maintenance reasons, and their use is assumed here.

The Bosch FMS transfer system[6] is taken as the basis for estimating conveyor costs. The Bosch system is suitable for manual or automatic or mixed assembly systems. Conveyors of almost any size and complexity can be constructed by combining items from the range of standardised system modules.

A rectangular conveyor layout (as in Fig. 1) is the lowest cost design. Other arrangements offer no distinct advantages. For automatic unloading of assemblies, the longitudinal belts are as closely spaced as possible. For manual unloading, the longitudinal belts are spaced 610mm apart, and connected at each end by powered transverse conveyors; robot safety guarding in this gap protects the unloading operator working on the return belt.

At each workstation location is a lift and indexing unit, which locates each arriving workcarrier within 0.1mm. This provision is essential for close-tolerance automatic assembly since workcarriers are not precisely guided on a moving belt. Buffer space for workcarriers must be provided at each workstation but usually normal workstation spacing is adequate in this respect.

A workstation may also be provided with code-reader and code-setter units, to read or set code memories on the workcarriers. Code memories control the movement of workcarriers through the system. For example, a code memory can be set at one workstation where an operation fails and read at later workstations, whose operations are thus inhibited.

Workcarriers are available in four lengths (160, 240, 320 and 400mm), and three widths (160, 240 and 320mm). If code memories are required, up to six may be fitted to each workcarrier. Tooling on each workcarrier to hold the assembly must also be provided; this will usually be specific to the product and may be the most costly part of the workcarrier.

Workcarrier size and number of code memories effectively fix the entire conveyor system design and cost. Conveyor cost is usually 20–30% of the cost of the robots in the system; large or high output assembly systems have higher conveyor costs.

Space and maintenance costs

Space and maintenance costs are frequently difficult to ascertain, but make a small but quite significant contribution to overall assembly costs.

The approach adopted here is to charge annual space costs on a floor area basis. If factory space is available for the assembly system, a charge of about £70 per square metre per year for heat, light, power and building costs should cover average conditions. If new space has to be built, the marginal cost may be mugh higher.

Annual maintenance costs are usually 8–10% of the total capital cost of the assembly system.

Space and maintenance charges must be recovered by apportioning them over the annual output of the assembly system. Typically, they can amount to about 18% of the cost of assembly.

Robotic assembly system manpower

An automatic assembly system is intended to function without operators. In practice, this goal is unattainable. Lotter[3] quotes a study on 21 automatic assembly lines in Germany which showed that the average running period was only 1.6min, with an average interruption time of 30s. 80% of interruptions are due to defective or dirty components or foreign bodies jamming parts feeding equipment.

Operator intervention is needed to clear jammed parts and restart the equipment. Since stoppages due to the causes indicated occur at random intervals, the system needs continuous operator supervision.

Operator intervention is also required to replenish parts feeders as they empty with component usage. With automatic signalling of a low component level to the operator, machine stoppage need not occur.

Though the assembly system itself is automatic, finished assemblies may be manually unloaded. This will be seen to be entirely practical, provided that the unloading operator can also rectify stoppages or replenish feeders.

Manpower for setting-up prior to operation is assumed to be covered by a suitable overhead rate on the operating manpower, as is supervision.

Results and conclusions

Cost of manual assembly

Using the same cost elements for manual assembly as for robotic assembly, assembly cost per part assembled is about 1.6 pence, irrespective of volume of production or size of assembly. This figure assumes a typical manual assembly time of 10s per part.

It is unrealistic to expect robotic assembly to replace manual assembly if no cost saving results, in view of the technical and business risks inherent in a new capital-intensive production process. Cost savings of at least 25% appear to be necessary to justify installing robotic assembly. Robotic assembly cost must therefore be less than $0.75 \times 1.6 = 1.2$ pence per part, for the process to be 'economic'.

Given a cost saving, other benefits such as more consistent product quality and less dependence on uncertain labour availability then become admissable as justification for assembly automation.

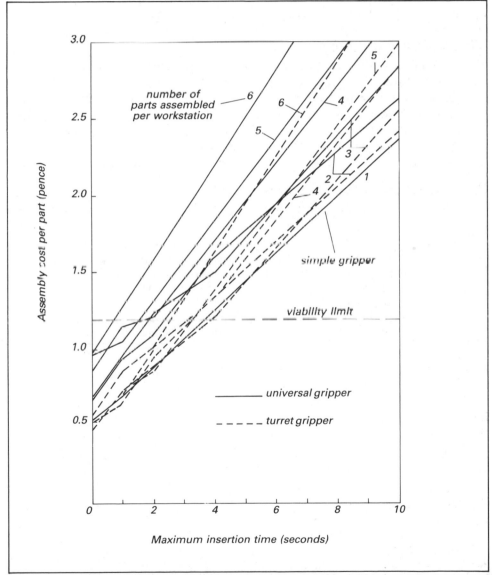

Fig. 6   Assembly cost and insertion time. Note, Fanuc A-Model 3 robot; 15 parts; two shifts over two year capital recovery

Insertion times

The time required to insert a component into the assembly very strongly influences the economic success of robotic assembly. Strictly, this is a product design consideration, since insertion time is determined by engagement lengths, clearances, tolerances, and so on. It should be noticed particularly that one long insertion time alone is sufficient to jeopardise economic viability if the assembly line cannot be balanced.

Fig. 6 shows how assembly cost varies with insertion time. In two-shift operation, with a two-year capital recovery period, robotic assembly will not be economic if insertion time exceeds 4s. Three-shift working or longer capital recovery periods ease this constraint. Robots fitted with turret grippers are distinctly less sensitive to insertion time.

Choice of robot and gripper

Fig. 7 shows assembly cost per part versus output, for the four Fanuc robots detailed in Table 2, each fitted with the various types of gripper. The clear trends to emerge are that turret grippers generally give slightly higher output and lower assembly cost, particularly as the number of parts assembled per workstation increases. A specially attractive aspect of turret grippers is that assembly cost may remain nearly constant over a wide output range. By choosing the number of parts assembled per workstation appropriately, the

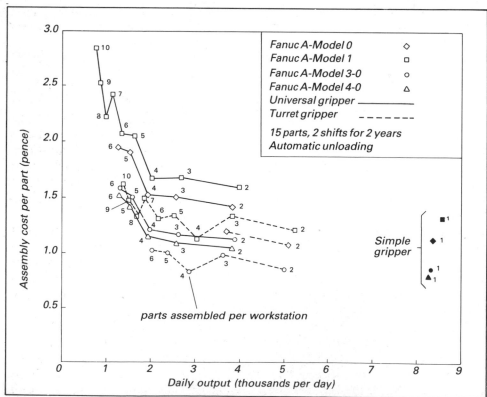

Fig. 7   Assembly cost and output for different robots and grippers

system can be tailored to a particular output. The robots, however, must have a large enough payload to carry a full turret. In contrast, assembly cost with universal grippers always increases as the number of parts assembled per workstation is increased and system output falls.

A simple gripper assembling one part per workstation always gives the least assembly cost and highest output. It is questionable, however, whether such a system would be more economic than a dedicated automatic assembly machine, since pick-and-place devices would be much cheaper than robots, and conveyor and feeder costs are the same for both dedicated and robotic systems. The robotic system would be favoured only if the task flexibility of robots has any value.

As might be expected, assembly cost increases with robot cost, when the different robots are all fitted with the same kind of gripper. A more expensive robot fitted with a turret gripper may, however, offer lower assembly cost than a cheaper robot fitted with a universal gripper, both grippers costing the same.

Choice of unloading method

In general, assembly cost is lower if the last robot in the line unloads finished assemblies automatically, except sometimes when the last robot has the sole task of unloading. The cost advantage of automatic unloading is small, however, being usually less than 5%. Manual unloading may well be more cost-effective if the unloading operator inspects finished assemblies, or performs an assembly task, as part of unloading. Either is possible since the time available for unloading is normally greater than 15s.

Overall system operation

At current robot cost levels, single shift operation will be uneconomic for ordinary commercial production, whatever the assembly, however the robotic assembly system is configured, and irrespective of any reasonable capital recovery period. For two-shift operation, some system configurations are economic even with a two-year capital recovery period, though small assemblies (about five parts) are unsuitable subjects for robotic assembly. Most system configurations are economic in three-shift operation with a two-year capital recovery period.

As the number of parts in the assembly increases, assembly cost per part falls slightly, and the variety of viable system configurations increases. The capital cost of viable systems also increases with part count; the minimum capital cost of a system for a 10-part assembly would be about £100,000, and for a 50-part assembly, about £250,000. The minimum investment for any viable system is unlikely to be much less than £100,000.

Investments of these magnitudes are economic only if the production volume is sufficiently large, and can be sold. The range of economic daily output, for 240 working days per year, is between 1250 and 5500 assemblies per day. This output could be achieved with between one and three operators per shift, the required number depending on output and assembly system size.

Acknowledgements

Research by The Hatfield Polytechnic into flexible automated assembly is supported by the SERC and by Delco Products Overseas Corp. This support is gratefully acknowledged.

References

[1] Boothroyd, G., Poli, C. and Murch L.E. 1982. *Automatic Assembly*. Marcel Dekker, New York.
[2] Miller, J. 1984. Robotic assembly line analysis. Hatfield Polytechnic, unpublished.
[3] Bracken, F.L. 1984. Parts classification and gripper design for automatic handling and assembly. In, *Proc. 5th Int. Conf.on Assembly Automation*, pp. 181-190. IFS (Publications) Ltd, Bedford, UK.
[4] Skoog, H. and Holmqvist, U. 1983. Matching the equipment to the job. *Assembly Automation*, 3(4): 211-214.
[5] Browne, A. 1984. A trainable component feeder. In, *Proc. 5th Int. Conf. on Assembly Automation*, pp. 85-93. IFS (Publications) Ltd, Bedford, UK.
[6] *Transfer Systems with Bosch FMS for the Assembly and Interlinking Technique*. Robert Bosch GmbH, Stuttgart, 1982. (Marwin Production Machines, Wolverhampton).
[7] Lotter, B. 1984. Basic requirements for unmanned automatic assembly in the field of precision engineering industry. In, *Proc. 5th Int. Conf. on Assembly Automation*, pp.287-298. IFS (Publications) Ltd, Bedford, UK.

Appendix – Cost analysis data

The following data is assumed throughout the paper:

- 1,2 or 3 shifts of 8 hours, 5 days per week, 240 days per year. No overtime.
- System down-time of 30min per day.
- Operator absence of 30min per shift.
- Assembly operator employment cost: £8000 per year. Overtime premium on 2nd and 3rd shifts: 25%.
- Tool designer and toolmaker rate: £10 per hour.
- Cost of space: £70 per square metre.
- System maintenance: 10% of installed capital per year.
- Largest part size: 50mm.
- Maximum and minimum part weights: 500 and 20g.
- Maximum and minimum total Z-axis travel: 100 and 50mm.
- Incidence of defective parts causing stoppages: 1%, assumed uniform.
- Operator time to rectify a stoppage: 30s.
- Operator time to replenish parts feeder: 150s.
- Operator time to manually unload finished assembly: 6s.
- Capacity of bowl feeders: 200 parts.
- Workcarrier. Size 240mm square. 1 code memory. Cost with tooling: £189.
- Conveyor costs: £7292 (manual unload) or £6028 (automatic unload) plus £1253 per workstation plus £203 per metre of system length plus 12%.
- Simple gripper. Total cost: £500. Grasp/release time: 0.3s.
- Universal gripper. Cost: £3000 plus £100 tooling cost for each part handled. Maximum and minimum grasp/release times: 0.8s and 0.3s. Weight 1kg.
- Turret gripper. Cost: £2500 plus £250 actuator and tooling cost per station. Grasp/release time: 0.3s. Weight: 1kg plus 220g per station.

ECONOMICS OF GENERAL-PURPOSE ASSEMBLY ROBOTS

G. Boothroyd
University of Massachusetts, USA

First presented in *Annals of the CIRP*, Vol. 31, No. 1, 1984. Reproduced by permission of the author and CIRP.

A study of the comparative economics of various automatic assembly systems using general-purpose assembly robots is described. In the study, usc is made of a typical candidate assembly whose profile is obtained from statistical analysis of actual products submitted by interested companies. It is shown that the choice of assembly system and the various parts presentation methods depends on three main factors: the annual production volume per shift, the number of parts in the product and the equipment payback period.

Research on the economic application of robots to automatic assembly has been underway at the University of Massachusetts for the past two years. Through working closely with various key manufacturing organisations, it has been possible to determine the kinds of assemblies that are generally considered candidates for robot assembly. It is found that these assemblies have quite similar characteristics and that a typical profile can be deduced for a candidate assembly. Using this typical candidate assembly, this paper considers the economics of some configurations of robot assembly systems of interest at present. The study will be restricted to those situations where general-purpose assembly robots can be applied and three basic systems will be modelled:

- A single-station machine with one fixture and one robot arm.
- A single-station machine with one fixture and two robot arms.
- A multi-station free-transfer machine with single robot arms at the various stations.

Parts presentation

When considering the suitability of products for robot assembly, it is important to consider the available methods of parts presentation. These range from 'bin-picking' using the most sophisticated vision systems, to the manual loading of parts into pallets or trays.

Much research is presently underway on the subject of programmable feeders and the impression is given that these can be used for systems that can assemble a variety of products. However, investigations have shown that most, if not all, of the 'programmable' assembly systems presently being developed in the USA will be devoted to the assembly of one product or one family of products. Thus, these systems will be flexible in the sense that they can be adjusted quickly to accommodate different members of the product family (different styles) or to accommodate product design changes. Nevertheless, the equipment used on these systems will be 'dedicated' to the product family regardless of whether it is 'special-purpose' equipment or 'programmable' equipment.

When the economic choice of parts presentation method is to be determined there will usually be only two basic types to consider:

- Those involving manual handling and loading of individual parts into part trays, pallets or magazines.
- Those involving automatic feeding and orientating of parts from bulk.

In some cases it is possible to obtain parts premagazined. Since these magazines will usually have been filled by hand or automatically from bulk, then the cost of the magazining will be borne by the supplier and will be included in the cost of the parts. Thus, the economics of using premagazined parts will have to be considered on a case-by-case basis with a knowledge of the increased parts cost.

For the present purposes, no distinction will be made in different types of automatic equipment because the cost of feeding each part automatically, CSA, will be given by multiplying the rate for the equipment (cents per second) by the time interval between which parts are required. Hence,

$$CSA = CF \times CR \times TAT \tag{1}$$

where CSA is the cost of feeding one part (cents); CF is the cost of the feeder for one part type including tooling, engineering, debugging, etc. (k\$); and TAT is the average station cycle time (seconds).

The equipment rate CR can be estimated from:

$$CR = 0.014/PS \qquad \text{(cents/k\$s)} \tag{2}$$

where PS is the number of shift-years worked during the equipment payback period.

Thus Eqn. (1) becomes:

$$CSA = 0.014 \times CF \times TAT/PS \tag{3}$$

The cost CSM of using a manually loaded magazine to present one part will be given by:

$$CSM = CM \times CR \times TAT + TM \times OP \tag{4}$$

where CM is the cost of the magazines (k\$), TM is the manual handling and insertion time for loading one part into the magazines (s) and OP is the operator rate (cents/s).

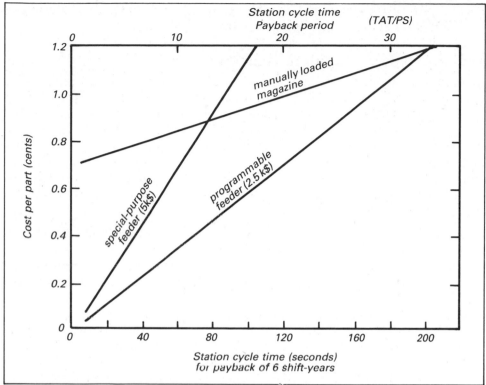

Fig. 1 *Cost of part presentation*

For a simple part the *Design for Assembly handbook*[1] gives a value of *TM* = 1.75s. A typical value of *OP* is 0.4 cents/s and thus Eqn. (4) becomes:

$$CSM = 0.014 \times CM \times TAT/PS + 0.7 \qquad \text{(cents)} \qquad (5)$$

Regarding equipment costs, these are likely to be much lower for a programmable feeder[2] than for a dedicated vibratory bowl feeder. This is because programmable feeders will require minimal tooling costs and setup, whereas dedicated vibratory bowl feeders require substantial tooling costs and setup, including the addition of a delivery track. It can be assumed that a dedicated vibratory bowl feeder, when supplied with simple tooling, engineered, debugged and fitted with the necessary delivery track, costs between $2000 and $7000[3]. With these figures in mind and remembering that no delivery track is required, a programmable feeder should cost between $1000 and $3000 per part type to be handled. For the purposes of initial comparison we will assume that a special-purpose feeder will cost $5000 and a programmable feeder will cost $2500.

Finally, if we assume that the cost of a special-purpose magazine or feed track is 1k$ and that the payback period for equipment is 6 shift-years then the comparison of feeding costs shown in Fig. 1 is obtained.

It can be seen that if a programmable feeder can be used then such a feeder will be the preferred method of part presentation for station cycle times less than about 200s. If a programmable feeder cannot be used then the special-purpose feeder would be more economic than the manual loading of magazines for station cycle times less than about 75s.

Profile of typical candidate assembly

Companies known to be interested in robotic assembly were requested to submit examples of assemblies that they considered were possible candidates for robotic assembly. Non-disclosure agreements were signed in order that up-to-date examples could be provided. These assemblies were then analysed for the purposes of up-dating the 'product design for robot assembly' system being developed by the author. However, similarities in the nature of the assemblies led to the idea that a 'profile' could be deduced for a typical candidate assembly. Analysis of 12 assemblies indicated that if NP were the total number of parts, then:

- $0.78 \times NP$ were different part types. This would approximate to the case, for example, where in an assembly containing ten parts, three were identical. Information such as this is important because, for example, in a single-station machine using one arm, these three parts could all be fed by the same feeder.
- $0.22 \times NP$ parts could be assembled automatically but not fed automatically and required manual handling. Such parts would be manually loaded into pallets, magazines or feed tracks.
- $0.32 \times NP$ parts could be fed and orientated using only $0.26 \times NP$ 'programmable' feeders because some of the parts are identical. The word programmable is used here to describe the type of feeder, presently under development, which can quickly be tooled for a particular part, thus reducing tooling and engineering costs.
- $0.39 \times NP$ parts could be fed and orientated using $0.24 \times NP$ special-purpose vibratory bowl feeders.
- $0.12 \times NP$ parts required manual handling and insertion or some manual manipulation. On a transfer machine, these would require a separate manual station. On a single-station system the partially completed assemblies would be delivered to a manual station external to the system and then returned to the system after completion of the manual operation. Thus, two pick-and-place operations involving a robot arm would be required in addition to the manual operation.
- $0.18 \times NP$ parts required special grippers or tools in addition to the basic grippers.
- $0.15 \times NP$ parts required that the assembly be transferred to a special-purpose workhead or tool to complete the required operation.
- The average assembly time per part for a single-station assembly system with one fixture and two robot arms was 3.29s; for a system with one arm it was 5.4s.
- If the product were to be manually assembled, the average assembly time per part would be 8.15s.

Using these figures it is possible to estimate the cost of assembly using different configurations of robot assembly systems and compare the results with the cost of manual assembly.

Table 1 Breakdown of equipment costs for a one-arm single-station system to assemble NP total parts

Item	Number required	Cost per unit (k$)	Total cost (k$)
Robot arm, sensors, controls, etc.	1	100	100
Workfixture	1	5	5
Standard grippers	1	5	5
Special grippers or tools	$0.18 \times NP$	5	$0.9 \times NP$
Magazines, pallets, etc.	$0.22 \times NP$	1	$0.22 \times NP$
Programmable feeders	$0.26 \times NP$	2.5	$0.65 \times NP$
Special-purpose feeders	$0.24 \times NP$	5	$1.2 \times NP$
Manual stations and associated feed tracks	$0.12 \times NP$	5	$0.6 \times NP$
Special-purpose workheads	$0.15 \times NP$	10	$1.5 \times NP$

Single-station systems

Equipment costs

First considering a single-station system with one work fixture and one arm, the costs of the assembly equipment for an assembly containing NP parts are listed in Table 1.

Summing these costs gives the total system cost:

$$CET(1) = 110 + 5.07 \times NP \tag{6}$$

Similarly the cost for a single-station system with one workfixture and two robot arms is found to be:

$$CET(2) = 165 + 5.97 \times NP \tag{7}$$

The first term has increased because of the additional robot arm and gripper. The second term has increased because, with a two-arm system, in order to keep the cycle time low it has been found best to share repeated operations between the two arms and duplicate the parts feeders.

Operator costs

One of the major problems when modelling automatic assembly systems is how to estimate the cost associated with the manual operator responsible for tending the machine and clearing stoppages due to faulty parts. On a multi-station high-speed assembly machine faults can occur so frequently that a full-time machine operator may be required. This same operator can also maintain adequate supplies of parts in the various automatic feeders. Under these circumstances, assemblies are produced at high rates and the cost of this operator forms a relatively small portion of the total cost of assembly. However, with low-speed single-station machines using robot arms, the cost of a full-time operator to tend the machine can comprise more than 50% of the cost of assembly.

For the purpose of the present analyses, estimates will be made of the actual manual time involved in producing each assembly. Where parts are loaded into magazines or pallets, or must be manually inserted, the total time required will be estimated. It will be assumed that if this total manual time is less than the system cycle time then the operator can be engaged on other tasks unrelated to the particular assembly machine being modelled. Similarly, the time necessary for tending the machine will be estimated and it will be assumed that the machine operator can be fully occupied by tending several such machines.

It should be emphasised that these assumptions will give the most optimistic estimates of assembly costs especially while robot assembly is in the development stage. However, it can be argued that unless machines are eventually designed that can run with the minimum of attention, then they will simply not provide an economic alternative to manual assembly. In fact, if a single-station general-purpose assembly machine were to require one full-time operator then give assistance in the form of fixtures, etc., that same operator could probably perform the assembly tasks at a similar rate to that of the robot!

Part quality

Machine breakdowns due to faulty parts, have two effects on the economics of the system. First, the average time to produce one assembly is greater than the system cycle time because of down time. Secondly, during a breakdown, a machine operator will be needed to clear the fault and restart the system.

If the average time taken to clear a fault and restart the system is TD, the number of parts in the assembly is NP and the average proportions of defective parts that will cause a fault is XP, then the average down time per assembly TDA will be given by:

$$TDA = XP \times NP \times TD \tag{8}$$

Basic cost equation

The basic equation for the total assembly cost CST per assembly is now given by:

$$CST = (TAT + TDA) \times CET \times CR + (TOT + TDA) \times OP \tag{9}$$
$$\underbrace{\qquad\qquad\qquad\qquad}_{\textit{equipment costs}} \qquad \underbrace{\qquad\qquad\qquad}_{\textit{operator costs}}$$

where TAT is the station cycle time (neglecting the down time due to defective parts); TDA is the system down time per assembly due to faulty parts, given by Eqn. (8) – it will be assumed that the ratio of faulty parts is 0.01 and the down time for each fault is 60s so that TDA is given by 0.6 × NP; CET is the total cost of equipment in k\$ (including debugging, etc.), given by Eqns. (6) and (7); CR is the basic equipment rate measured in cents per second per 1k\$ of equipment – for a payback period of PS shift-years CR is equal to 0.014/PS and is measured in cents/k\$s; TOT is the operator time per assembly for loading the magazines and performing the required assembly operations – in the present example the time taken to load one part into a magazine or to perform one assembly operation can be estimated from the *Design for Assembly Handbook*[1] to be 1.75s, thus using the figure for proportion of parts requiring manual handling or assembly given earlier,

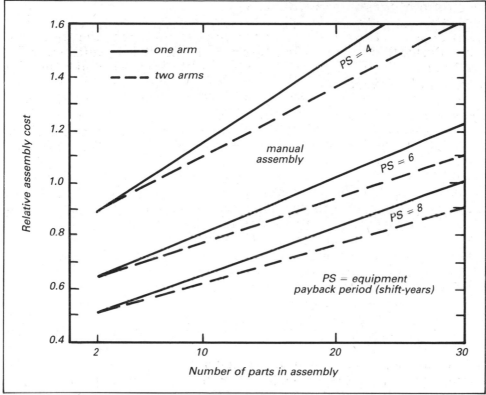

Fig. 2   Effect of number of parts on relative assembly cost for single-station systems

TOT is 0.6 × *NP*; and finally *OP* is the operator rate which is approximately 0.4 cents per second depending on wage rates and overheads.

Substitution of all these values and the equipment costs determined earlier into Eqn. (9) gives the following equations for the assembly cost per part.

For a one-arm system:

$$CST(1)/NP = (8.84 + 0.41 \times NP)/PS + 0.48 \qquad (10)$$

and for a two-arm system:

$$CST(2)/NP = (8.99 + 0.32 \times NP/PS + 0.48 \qquad (11)$$

After dividing by the average manual assembly cost per part of 3.26 cents, Eqns. (10) and (11) are plotted in Fig. 2 where it can be seen that the economics of one-arm and two-arm single-station systems are very similar. Therefore, the decision on which type of system to use must be made on an individual basis. However, it can be seen that with a payback period of 6 shift-years for example, single-station automatic systems used for assemblies containing more than about 22 parts are unlikely to be economic. An average station cycle time for such an assembly would be approximately 113s for a single-arm system and 72s for a two-arm system.

Multi-station transfer systems

Equipment costs

When the demand for a product is greater than could be assembled on a single-station assembly machine, duplicates of such machines could be installed, or alternatively a multi-station free-transfer machine could be employed. Our typical candidate assembly profile can again be used to study the economics involved in the latter choice.

On a free-transfer machine it is usually not possible to perform manual operations at any station where automatic operations are carried out. Similarly, those operations involving special-purpose workheads must be carried out at separate stations. From the information on the profile of the candidate assembly, the number of manual stations will be $0.12 \times NP$ and the number of stations with special-purpose workheads will be $0.15 \times NP$. The remaining parts ($0.7 \times NP$) will be assembled by robots and the number of stations will depend on the required assembly rate. If we assume 5 million seconds per shift-year of useful working and the assembly volume per shift is VS (million) then the average station cycle time (neglecting down time) would be $5/VS$. The figure for the average assembly time per part for a one-arm system used earlier was 5.14s, so the maximum number of parts assembled at one robot station NR would be given by:

$$NR = 5/(VS \times 5.14) = 0.97/VS \qquad (12)$$

and the number of robot stations NT on the machine would then be:

$$NT = 0.7 \times NP/NR = 0.72 \times NP \times VS \qquad (13)$$

Table 2 gives a breakdown of the equipment costs for a multi-station free-transfer machine. Summing these equipment costs gives the total cost $CET(T)$ for a free-transfer machine:

$$CET(T) = (11.4 + 78.5 \times VS) \times NP \qquad (14)$$

It should be noted that this equation is valid only for assemblies containing at least six parts and where the station cycle time is at least 8.15s to allow for a typical manual assembly operation. This latter figure corresponds to an annual production volume per shift VS of about 0.6 million.

Table 2   Breakdown of equipment costs for a multi-station free-transfer machine to assemble NP parts at a rate of VS million assemblies per shift

Item	Number required	Cost per unit (k$)	Total cost (k$)
Robot arms, sensors, controls, etc.	$0.72 \times NP \times VS$	75	$54 \times NP \times VS$
Transfer device and 3 workcarriers for each workstation	$0.72 \times NP \times VS$ $+ 0.27 NP$	29	$20.9 \times NP \times VS$ $+ 7.8 \times NP$
Grippers	$0.72 \times NP \times VS$	5	$3.6 \times NP \times VS$
Magazines, pallets, etc. for manual loading	$0.22 \times NP$	1	$0.22 \times NP$
Programmable feeders	$0.26 \times NP$	2.5	$0.65 \times NP$
Special-purpose feeders	$0.39 \times NP$	5	$1.2 \times NP$
Special-purpose workheads	$0.15 \times NP$	10	$1.5 \times NP$

Basic cost equation

As with the analysis for single-station machines, it will be assumed that manual operators are required only for the direct assembly and fault correction procedures and can be engaged on tasks unrelated to the particular assembly for the remaining time. The down time on a properly designed free-transfer machine approaches that for one individual station. Thus, the number of parts NP in Eqn. (8) should be regarded as the number of parts NA assembled at one robot station and given by Eqn. (12).

For a free-transfer machine,

$$TDA = 0.58/VS \tag{15}$$

and the station cycle time is:

$$TAT = 5/VS \tag{16}$$

Substitutions fof Eqns. (14), (15) and (16) in Eqn. (9) gives:

$$CST(T)/NP = (6.13 + 0.89/VS)/PS + 0.23/(NP \times VS) + 0.24 \tag{17}$$

After dividing by the manual assembly assembly cost per part, Eqn. (17) is plotted in Fig. 3 to show the effects of annual production volume and station cycle time on the cost of assembly. For comparison purposes, the results for single-station machines are shown on the same figure.

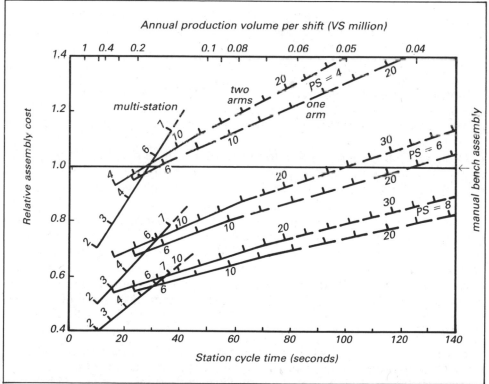

Fig. 3   Effect of payback period station cycle time and number of parts in assembly on relative assembly cost for various robot assembly systems (PS = payback period in shift-years, number of robot assembly operations per station shown)

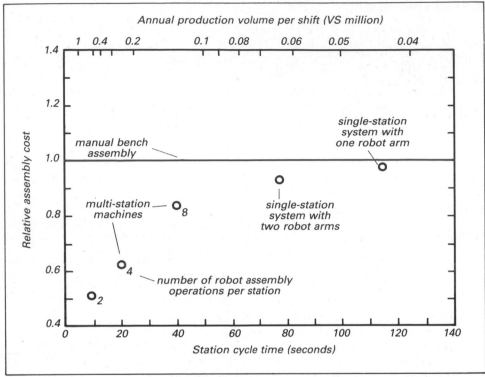

Fig. 4 *Alternative robot assembly systems for a typical assembly containing 20 parts (payback period in shift-years, PS = 6)*

Suppose a typical assembly consists of 20 parts; two requiring manual assembly and two requiring special-purpose workheads. Fig. 3 can be used to generate the alternative shown in Fig. 4 for a typical payback period of 6 shift-years.

It can be seen from Fig. 4 that for an annual volume per shift VS of around 45,000 the one-arm station system would be preferred. If VS is around 65,000 then a two-arm single-station system would be chosen. For values of VS between 160,000 and 500,000 per year multi-station machines would be the most economic. If it were possible to divide the 20 part assembly into two similar subassemblies of ten parts each, then it can be deduced from Fig. 2 that the preferred arrangements would be two one-arm single-station systems for annual volumes of 90,000 and two two-arm single-station systems for around 150,000. However, it should be remembered that the cost of mating the two smaller assemblies must be taken into account.

Concluding remarks

It can be seen from the work described here that for a particular assembly and for a given equipment payback period the choice of robot assembly system depends mainly on the annual production volume. When the production volume is less than about 40,000 per shift, robot assembly is unlikely to be economic for a typical payback period of 6 shift-years. For assemblies containing 20 parts or less required at annual volumes between 40,000 and 200,000 per shift the choice of a one-arm or two-arm assembly system would depend on the volume required. For higher volumes and larger assemblies multi-station machines should be considered with the provision that no more than two to six assembly operations are arranged at each robot assembly station.

Acknowledgements

The author wishes to express his appreciation to those graduate students involved in the design for robot assembly project. These students are P. Carter, Y. Ho, L. Hu, J. John, R. Langmoen, C. Lennartz, and S. Ro. In particular the help of C. Lennartz with the economics of parts feeding is gratefully acknowledged. The author also wishes to thank Dr. A.H. Redford and Dr. P. Dewhurst for reading the paper so carefully and for their numerous helpful suggestions.

The work described in this paper was carried out with the aid of grants from the National Science Foundation and from the following companies: AMP Inc., Digital Equipment Corporation, General Electric Company, IBM Corporation, Westinghouse Electric Corporation and Xerox Corporation.

References

[1] Boothroyd, G. and Dewhurst, P. 1983. *Design for Assembly Handbook.* University of Massachusetts, Amherst, MA, USA.
[2] Zenger, D. 1983. *Passive Orienting Systems for Double Belt Feeders.* M.S. Project Report, Department of Mechanical Engineering, University of Massachusetts, Amherst, MA, USA.
[3] Lynch, P.M. 1976. *Economic-Technological Modeling and Design Criteria for Programmable Assembly Machines.* Charles Stark Draper Laboratory, Inc., Cambridge, MA, USA.

Lennar
or al-
aper

COST ANALYSIS FOR MULTI-ARM ROBOTIC ASSEMBLY

A.H. Redford, E.K. Lo and P. Killeen
University of Salford, UK

First presented in *Assembly Automation* (November 1983). Reproduced by permission of the authors.

Unlike all other forms of automatic assembly equipment, the multi-arm assembly robot assembles parts serially rather than in parallel, and this leads to comparatively long cycle times for the equipment. Automatic parts feeders which are suitable for high-speed assembly are, therefore, under-utilised when used in this application. Analyses which have been developed to determine the cost of assembly using a multi-arm assembly robot fed by a wide variety of different feeding systems and for a wide range of product styles, product mixes and batch sizes, are described. The high-cost elements of the various feeding systems are identified and a specification for a feeding system which would be appropriate for robotic assembly is outlined.

Of the various types of programmable assembly, robotic assembly differs from all other forms in that the operations are carried out using two or more robot arms with the assistance of a limited number of auxiliary toolings, and in many cases, without the use of any transfer mechanism. Very often it is organised in cells, designed for maximum flexibility, and built to cope with the assembly of a variety of product styles manufactured in relatively small batches at low speed. Despite its immense potential benefits, assembly by multi-arm robots has failed to make a significant impact commercially for a variety of reasons, notably the high capital cost of the robot, the inflexibility of its peripherals, and the high cost of part presentation[1].

Traditionally automatic part feeding and orientation is carried out by a feeding device which is configured to meet the requirements of a particular or, at the most, a very restricted range of part types. Although this dedicated hardware causes few problems to hard automation where production runs are very long, changes in design are rare, and where production rates are high, it is often too rigid for robotic assembly. Not only does the latter demand programmable or interchangeable peripherals, but its relatively long cycle time and therefore low part consumption rate has made the cost of automatically feeding parts to a robot disproportionately high and expensive enough in many cases to make manual feeding justified.

Various attempts have been made to increase the versatility of automatic part feeders. One popular concept is to restrict the feeder's ability to deal with part presentation, whilst leaving part recognition to a vision system and the subsequent orientation to the robot [2]. Another approach is to enhance the versatility of a conventional feeder by incorporating programming capability to its orientating devices [3]. Both methods are technically feasible, but high equipment cost and low operating speed are the two major deterrents to their commercial application.

The objective of this work is to study the cost of assembly using a multi-arm robot and one or more of the various existing and hypothetical feeding systems operated under a range of circumstances. A mathematical model has been constructed to examine the economics of eleven feeding systems which are potentially suitable for robotic assembly; the proposed systems, the modelling algorithm, and their relative merits under various circumstances as depicted by the model are described. A short list of feeding alternatives which could be appropriate for robotic assembly is also given.

Parts feeding

Basically, parts can be fed either indirectly by magazines or directly via a feeder or an operator. In the former case the magazines can be loaded either manually, by the prior manufacturing operation, or by different types of feeders. Similarly, parts fed directly to a robot may be orientated by parts feeders, by the robot arm with the aid of a vision system, or manually. The possible alternative methods of feeding parts to a robot are outlined below (see also Fig. 1).

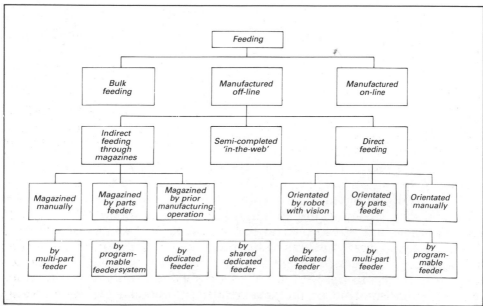

Fig. 1   Feeding method identification

Dedicated feeders

This system consists typically of vibratory feeders, each of which is equipped with one or more bowls dedicated to a particular part type. Parts common to different styles of the product will share the same bowl, whereas bowls used for different parts in an alternative style of the product will use the same drive unit. Thus, the number of bowls necessary for a planned variety of products will be equal to the number of distinct part types, whereas the number of drive units needed will be equal to the maximum number of parts in any style of the product.

Programmable feeders

A programmable feeder is considered to be one which, by the use of suitable automatically adjustable tooling, is versatile enough to feed and orientate a wide variety of parts, independent of their shapes, size, etc. Thus, the number of feeders required will be equal to the maximum number of part types for any of the product styles. The changeover time for the feeder will be sensibly zero, though the time to adjust or change ancillary equipment will still need to be taken into account.

Dedicated feeders serving more than one robot

As mentioned previously, dedicated feeders are under-utilised when used in robotic assembly applications, and it was thought that dedicated feeders serving more than one robot would result in more effective use of the feeder. Clearly, this type of system would only be practical if a number of multi-arm robots were being used in close proximity.

Feeders aided by vision

The vision system considered was modelled on equipment developed by Hitachi [4] which consists of a multi-level bowl feeder which is capable of delivering parts singly but not in orientation, a set of parts reversing units, a moving solid-state camera and the associated image-processing equipment. Parts are fed into the reversing units scanned by the camera and are either accepted or reversed and rescanned. Parts in a limited number of orientation, are then picked up by the robot and turned into the required orientation. Because the bowls do not need to be configured to perform parts orientating, they are easily adapted to accommodate different part types, but the range of part shapes capable of being handled by this type of system is limited.

Multi-part feeder

The multi-part feeder is a non-standard feeding system designed to feed and orientate several parts for presentation to a robot using multiple tracks but a common drive unit [5]. It takes the form of a linear vibratory unit equipped with multi-compartment storage containers and elevators (Fig. 2). Components from each compartment are raised to the linear tracks on which feeding and orientation are performed. In this study it was assumed that the feeder would be capable of handling two different part types.

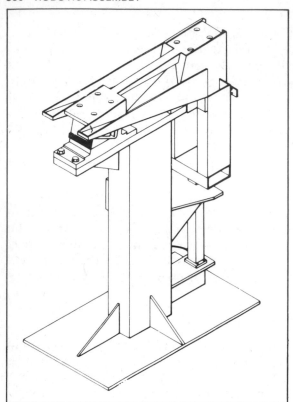

Fig. 2   Linear multi-part feeder

Magazine systems

Since automatic feeding systems for robotic assembly are invariably working at well below maximum rate, it is possible that some advantage could be gained by loading magazines using automatic parts feeders working at higher rates and then transferring the magazines to the robot's assembly points. Although the magazines would be expensive, this type of system has the further advantages of providing an inspection function and decoupling the assembly operation from the feeding. In the study, apart from magazines loaded by dedicated feeders, programmable feeders and multi-part feeders, two further magazine feeding systems, namely manual loading of magazines and magazines loaded at the point of piece-part manufacture, are also examined.

Manual feeding

As a reference for comparing the various feeding systems, manual loading of assembly robots was analysed where an operator would load parts into feed tracks in suitable locations in the robot's working area.

Costs of assembly

Generally, the cost of assembly consists of two elements: direct labour and depreciation of capital equipment. The former can be classified into material handling cost, system tending cost, and changeover cost, whereas the latter is made up of basic equipment depreciation and tooling depreciation.

Material handling cost

Material handling cost includes all labour cost incurred in the transportation of parts in bulk or in an organised manner in bins, magazines or racks from a specific stock point to various working areas and, where necessary, the transportation of empty containers from those areas back to the stock point. It can be expressed mathematically as:

$$C1 = FNC1(B, H, T1)$$

where $FNC1$ is the material handling cost function, B is the number of assembly batches, H is the number of handlings per batch, and $T1$ is the total time for each handling.

System tending cost

System tending refers to the basic labour requirement which would be essential for the day to day running of the robot and its feeders. Its cost $(C2)$ is a product of the assembly cycle time, the relevant labour hourly rate, and the number of operators involved. The latter depends on the ratio of the operator's working time to the machine running time and is a function of system design and layout rather than the complexity of the product. It is derived from:

$$W=FNW (t1, t2, t3, T2)$$

where FNW is the labour requirement function, $t1$ is the time for correcting robot fault, $t2$ is the time for correcting feeder stoppages, $t3$ the time for changing parts, and $T2$ is the allowed part assembly time.

As an assembly robot is usually far more expensive than its peripheral equipment, its operating rate and its average downtime determines the assembly cycle time. Feeder stoppages, although considered in the algorithms, would be unlikely to halt the assembly operation since orientated parts stored in the feed track create a buffer stock usually sufficiently large to meet the rate of parts consumption.

Changeover cost

Whenever a different style of product needs to be processed, quite often a certain amount of peripheral equipment adjustment or replacement would be necessary for both the feeding devices and the assembly fixtures. The changeover time required would depend on the type of feeding system being used, is a function of the number of changes needed to be made, and is subject to the sequence in which the batches are being scheduled. Alternatively, the labour rate is equal to the sum of the costs of both the machine setter and the system tender(s), since the latter is likely to be idle during the time of changing. In general, the cost incurred for a batch of product assembled according to a specific production schedule is:

$$C3 = FNC3 \ (T3, N, \ W1, W2)$$

where $FNC3$ is the changeover cost function, $T3$ the unit equipment changeover time, N the number of pieces of equipment involved, $W1$ the equipment setting rate, and $W2$ is the system tending rate.

Equipment depreciation cost

Equipment in this context refers to the part of the assembly system which is not specially designed to either feed or insert a particular component. This part of the equipment can be modified, with additional tooling, to accommodate a variety of products and parts and therefore is depreciated over a period of time rather than entirely against a particular job. It includes the purchase and commissioning costs of the relevant equipment, and is charged to the job according to the equipment occupying time. Thus,

$$C4 = FNC4 \ (C, \ T\emptyset, \ T)$$

where $FNC4$ is the equipment depreciation cost function, C the costs of assembly and feeding equipment, $T\emptyset$ the life of equipment, and T is the total production time of job.

Tooling depreciation cost

Tooling is that part of a feeding system which is dedicated to the feeding of a specific part. This would be redundant at the end of the job and therefore would be depreciated against the job. For a particular set of tooling,

$$C5 = C \times N$$

where C is the unit tooling cost and N is the number of sets of tooling required.

Modelling algorithms

In order to justify the use of any one of the feeding systems described, a mathematical model has been constructed to calculate the cost of feeding and inserting a family of products, each of which has its own product structure, demand pattern and batch sizes. The model was initially based on the assumption that one particular method of feeding is to be used for every part in the product. This has now been extended to include the practical situation where different means of feeding are used for the various part types in the product family. Despite the applicational difference, the modelling algorithms are basically the same for both situations and the same data base is applied. A few important assumptions that are made in developing the algorithms are discussed below.

Assembly cycle time

A number of factors determine the average length of an assembly cycle: the nature of the operation, the layout of the workplace, the capability of the robot and, perhaps most important of all, the complexity of the product. For simplicity, this model assumes that the cycle time for any assembly operation will only be a function of the complexity of the product, and the other factors will be accounted for by modifying the unit part assembly time (the average time required to handle and insert a component and, if necessary, to change the gripper). Hence,

$$S = P \times S\emptyset$$

where S is the time to assemble a complete product, $S\emptyset$ is the unit part assembly time, and P the number of constituents in the assembly.

Scheduling strategy

As the time involved in equipment changeovers depends very much on the frequency of changes made between two consecutive batches, the ways that the assembly batches are being organised will influence both the costs of changeover and equipment depreciation. Instead of defining a production schedule for every batch of the products, or simulating the effect by complete randomness, the model tackles the problem by providing a choice of one of three scheduling levels, i.e. least feeder changes between batches, frequent feeder changes between batches, and the mean of the two. It assumes feeding equipment assigned to common parts shared by every product within the family needs no changeover; whereas the rest may require minimum, maximum or moderate alternations, depending on other production constraints.

Dedicated feeders serving more than one robot

Assemblies produced using this system are performed in parallel by several robots with the longest batch assembly time taken to be the run time for the whole system. When the entire production lot has been completed, the robots are assigned to other jobs requiring other feeding equipment and, at this time, equipment changeovers will become necessary. Thus, the total lot assembly time depends on the combinations of different product assembly times, their batch sizes and the number of batches required. It is assumed that the total number of assembly runs is equal to the number of batches produced for the most frequently changed product (i.e. if three products are to be assembled and if the numbers of batches required are 3, 7 and 10, respectively, then the total number of assembly runs will be 10). The number of products being produced for each run will vary, whereas the length of each assembly run may change too and will be equal to the longest batch cycle time of the product being assembled.

Magazines loaded by feeders

Whilst using feeders to load magazines will increase feeder utilisation, and hence reduce equipment depreciation cost, it will also incur extra costs in tooling and changeovers. In order to optimise these costs, the model firstly fixes the feeder variable to the minimum (i.e. 1). It then computes the summations of changeover and magazine costs by assuming a range of magazining batch sizes and selects the lowest of these. The process is repeated by increasing the number of feeders, one at a time, until there is a feeder dedicated to each part type in the most complicated product. Comparing the lowest summations of all these alternatives will then give the most economic combination of feeders and magazines for the product family under consideration.

Implementation and testing

The mathematical model was implemented using BASIC-PLUS running under RSTS/E mounted on a Systime Series 6400 (DEC PDP11/44 processor). It consists of a database management program, two cost evaluation programs (the 'one system for all parts' approach and the 'mixed' approach), and two output analyses programs capable of listing, plotting and comparing results. Data is stored using BASIC-PLUS's Vitual files which provide a simple structure to record the product/constituents relationship. At present, a maximum of 60 operating parameters, 20 product styles and 100 part types can be stored, though these limits can easily be extended if necessary.

Test data

Although the model is designed to study the economic viability of the various feeding systems operated under different assembly environments, it would be useful if it can generate some simple rules which will indicate the trends in the current economic circumstances. After some careful considerations, most of the operating parameters (unit labour rate, equipment efficiency and reliability, equipment and tooling costs, etc.) were fixed using realistic values while selected variables such as the unit part assembly time $(2 - 3s)$, life of the feeding equipment (2, 3 and 5 years), and number of working shifts (2 and 3) were entered and their effects on feeding costs studied using the 'one system for all parts' model. As the result of these studies did not show any significant changes in the order of preference of the feeding systems, subsequent tests concentrated on the effects of product structure and demand pattern on the cost of parts feeding.

Annual demand

For equipment running on three shifts a day with a unit part assembly time equal to 3s, the number of parts that can be assembled annually will be approximately 6,480,000. Thus, a single robotic assembly system will be capable of producing 1,620,000, 648,000 or 324,000 products annually if each product consists of 4, 10 or 20 parts, respectively. Conversely, if there are 10

Table 1   Product structure of a family of four 4-parts products (test series I)

Product	Constituent P/N			
P1	PN1 ,	PN2 ,	PN3 ,	PN4
P2	PN1 ,	PN2 ,	PN3 ,	PN5
P3	PN1 ,	PN2 ,	PN3 ,	PN6
P4	PN1 ,	PN2 ,	PN3 ,	PN7

Table 2   Family sizes and requirements of one 4-parts product (test series I)

No. of products/family	No. of part types/family	Demand/product
1	4	1,620,000
4	7	405,000
8	11	202,500
12	15	135,000
16	19	101,250
20	23	81,000

product styles within a product family, each of which consists of 10 parts, then the average annual requirement per product in this family will be 64,800. Such an argument provides a common reference in relating the structure of a family of products to its demand pattern and is applied to the following hypothetical product families.

Test series I

Test series I simulates the type of production environment where the total demand per product family is too small to be justified for line assembly though the batch size of each product can be as large as its annual demand. It is assumed that the family consists either of 1, 2, 4, 8, 12, 16 or 20 products each of which may be composed of 4, 10 or 20 part types. Furthermore, each part type is only required once in a product and the structure of a product within a family differs from one another always by one and the same part. Tables 1 and 2 show the structure and requirements of products constructed for this series of tests.

Test series II

This simulates the situation where batch assembly is essential in the production of a family of products. Here, the annual demand of each product is to be met by equal batches, the smallest of which was considered to be 50. The size then increases gradually until batch size no longer has any significant effect on the order of preference of the individual feeding systems. Instead of assuming the one to one parts swapping criteria stated previously, this series works on the assumption that any one 'part' in a product may be of only one of two possible part types. Tables 3 and 4 illustrate the structure of a 4-parts product and the sizes and requirements of the families under study.

Table 3   Product structure of a family of four 4-parts products (test series II)

Product	Constituent P/N			
P1	PN1 ,	PN2 ,	PN3 ,	PN4
P2	PN1 ,	PN2 ,	PN3 ,	PN5
P3	PN1 ,	PN2 ,	PN6 ,	PN4
P4	PN1 ,	PN2 ,	PN6 ,	PN5

Table 4   Family sizes and requirements of one 4-parts product (test series II)

No. of products/family	No. of part types/family	Demand/product
2	5	810,000
4	6	405,000
8	7	202,500
16	8	101,250

Results

Test series I

Fig. 3 relates the cost of assembly per part to the number of products in a family for the various product sizes. As expected, assembly cost increases as the size of the product family increases, despite the fact that the total number of parts assembled throughout the entire lot remains constant. This is largely

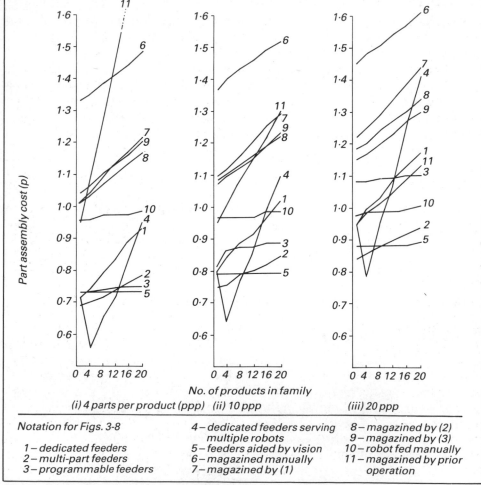

Fig. 3   Part assembly cost/product family size relationship (batch size = annual demand)

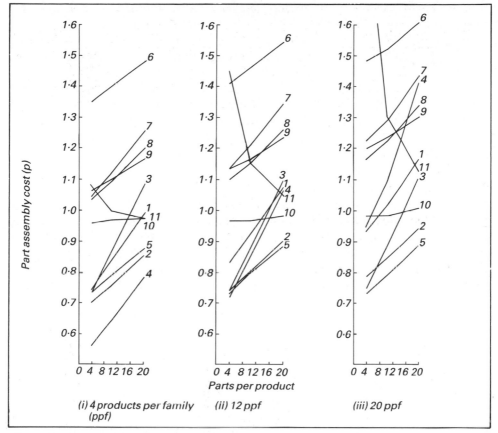

Fig. 4   Part assembly cost/product size relationship (batch size = actual demand)

due to the rise in changeover cost and, in some cases, tooling costs. However, the phenomenon does not apply to some systems such as programmable feeders, feeders with vision and manual loading which incur minimal changeovers between different products.

As the number of constituents in a product increases, the assembly cost for most systems also increases (Fig. 4). Magazines loaded by prior operation is an exception simply because its tooling cost, accounted largely by the number of magazines deployed to meet an hourly production, decreases significantly as the product's assembly cycle time increases.

Nevertheless, both figures reveal that under the defined production environment, magazining by whatever means is expensive. It is always more economic to feed the robot directly using feeders or even manually than via a magazine system. However, if magazines have to be used, loading them at the prior manufacturing operation should be considered unless the number of magazines involved becomes too many to be justified. Then a choice can be made between magazines loaded by a programmable feeder or by a multi-part feeder.

Among the various direct feeding methods the multi-part feeder and the feeders with vision system are the two most economic means of presenting parts to a multi-arm robot. Using the present modelling strategy, the programmable feeder is greatly under utilised and hence expensive, because its versatility is not fully made use of when feeding parts which are common to all products.

Although the arrangement of sharing dedicated feeders by more than one robot can be subjected to many practical limitations, the system, when properly used, can be very economical. With the assumed operating conditions, it is increasingly effective as the size of the product family reduces. This occurs primarily because of higher labour productivity.

Test series II

The typical effect of batch sizes on feeding cost is shown in Fig. 5. Here, batch size, expressed in terms of 'shifts' of 8 hours, exhibits little influence on the unit cost of parts feeding once it exceeds unity (for a product size of 10 parts, this is equivalent to 960 assemblies) The order of economic preference then becomes relatively constant and, for most systems, a breakeven quantity does not exist. In fact, the graph shows practically no incentive to

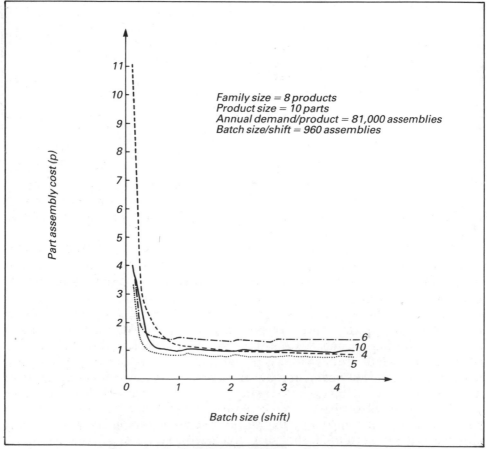

Fig. 5   *Part assembly cost/batch size relationship*

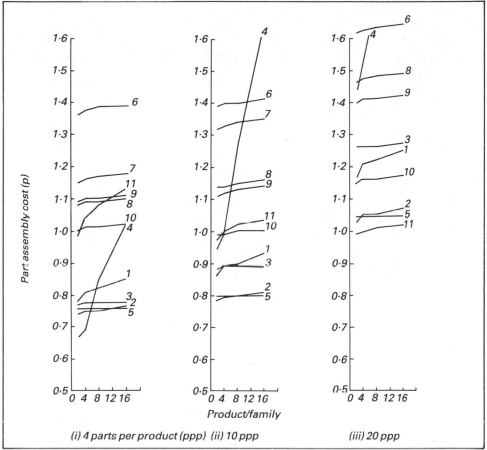

Fig. 6   Part assembly cost/product family size relationship (batch size = 1 product shift)

increase the batch size further than one shift, as the unit part assembly cost has already reduced to the level achieved by a batch size equal to the annual demand (in this case 81,000 assemblies).

As it is quite unusual to run an assembly batch in less than one shift, the assembly costs at this quantity are referred to in the following paragraph where the relationship between feeding costs and product structures is studied.

Figs. 6 and 7 show the effect of family size and product size on the costs of assembly, respectively. They are plotted in a similar way to Figs. 3 and 4 except, in this case, the batch size is equal to one shift of production instead of the annual demand. However, the findings are similar to those of the previous study, with magazine processes remaining relatively uncompetitive and the multi-part feeder and the feeders with vision system being consistently the most economic. It is further confirmed that large product size favours magazines loaded by the prior operation, whereas small family size has a positive effect on the choice of sharing dedicated feeders by more than one robot.

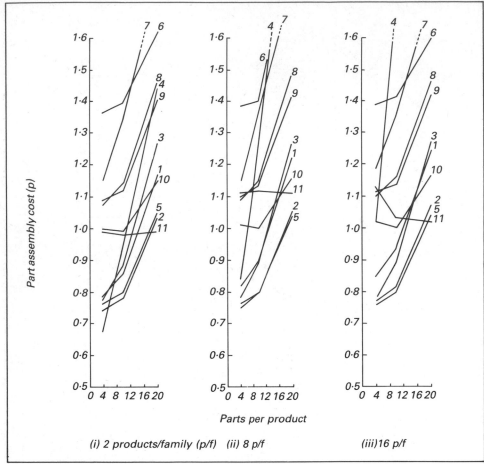

Fig.7   *Part assembly cost/parts per product (batch size = 1 product shift)*

Miscellaneous tests

As stated earlier, it is not to the advantage of the programmable feeder if it is applied to every part type which make up the family of products, irrespective of whether they are interchangeable or shared by one or more product in the family. To demonstrate this, the algorithm written for evaluating a mixed system of feeding methods was applied to a product family consisting of eight products with ten parts in each product and seventeen unique part types (test series I). It was assumed that two of the seventeen parts types have to be magazined, and that this would be done using the magazines loaded by the prior operation system, while the rest can be fed by one of the following means:

- All 15 parts using multi-part feeders.
- 7 parts which are common to all 8 products are fed by multi-part feeders, with the remaining 8 (unique and interchangeable) by a programmable feeder.
- All 15 parts fed using a feeder with vision.

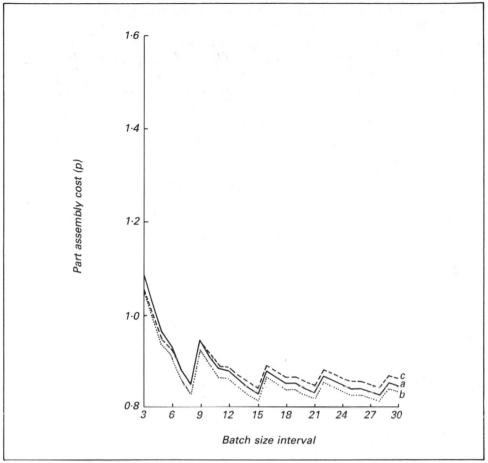

Fig. 8 Part assembly cost of a mixed system

It was found that of the three, the second is the cheapest (Fig. 8). Thus it may be concluded that provided the programmable feeder is allowed to use its versatility, it is the most appropriate equipment to use.

Conclusions

The main conclusions which can be drawn from the study so far are:

- For batch sizes above one shift of production, batch size becomes a second order effect, i.e. the rate of decrease of feeding cost with batch size is small. It does not decrease significantly even if the batch size is increased to the same level as the annual demand.
- For reasonable batch sizes, the differences between the best feeding system and the worst is surprisingly small (a 100% change).
- Compared with the average part assembly cost of a manual feeding and assembly system (typically 1p), robotic assembly can only be economically viable if the batch size is more than, say, 700 assemblies per shift and if the method of feeding is carefully chosen.

- Multi-part feeder, feeders with vision and programmable feeders are the feeding systems that are economical for robotic assembly. Technically they compliment each other but among them, only the multi-part feeder is a true substitute of the conventional vibratory bowl feeder, both in terms of technical capability and economic viability.
- With few exceptional circumstances, magazine feeding and sharing dedicated feeders by more than one robot are difficult to justify.
- Operating parameters, such as part assembly cycle time, period of depreciation and strategy of scheduling, affect the magnitude of feeding costs of the various systems but not the order of economic preference.

References

[1] Redford, A.H., Lo, E.K. and Killeen, P.J. 1982. Design for robotic assembly. In, *Proc. 3rd Int. Conf. on Assembly Automation*, 25-27 May 1982, Boeblingen, West Germany, pp.263-270. IFS (Publications) Ltd, Bedford, UK.
[2] Hill, J.W. 1980. Programmable bowl feeder design based on computer vision. *Assembly Automation*, 1 (1): 21-25.
[3] Cowart, N.A. et al. 1980. *Programmable Assembly Research Technology Transfer to Industry - Phase 2*, Year end report 1980. NSF Grant ISP 78-18773 (1981), pp.18-19, 63-67.
[4] Suzuki, T. and Kohns, M. 1981. The flexible parts feeder which helps a robot assemble automatically. *Assembly Automation*, 1(2): 86.
[5] Hollingum, J. 1982. Automated parts feeder design cuts cost of robot assembly. *Assembly Automation*, 2(4): 214-215.

Authors' Organisations and Addresses

P. T. Blenkinsop
PA Technology
Cambridge Laboratory
Melbourn
Royston SG8 6DP
England

C. Blume
Institute for Informatics III
Research Group: Process Control
    Computer Technology
University of Karlsruhe
7500 Karlsruhe 1
West Germany

G. Boothroyd
Department of Mechanical
    Engineering
University of Massachusetts
Amherst, MA 01003
USA

J. Browne
Department of Industrial
    Engineering
University College Galway
Ireland

A. Camera
DEA SpA
Corso Torino, 70
100024 Moncalieri
Torino
Italy

K. Collins
Cranfield Robotics and
    Automation Group
College of Manufacturing
Cranfield Institute of Technology
Bedford MK43 OAL
England

S. Ericsson
Saab-Scania
Scania Division
S-151 87 Södertälje
Sweden

J. W. Franklin
Automatix Inc.
1000 Tech Park Drive
Billerica, MA 01821
USA

G. Gini
Department of Electronics
Politecnico di Milano
Milan
Italy

J. J. Hill
Department of Electronic
    Engineering
University of Hull
Hull HU6 7BX
England

U. Holmqvist
ASEA Robotics
S-72183 Vasteras
Sweden

J. F. Laszcz
IBM Corporation
P. O. Box 12195
Research Triangle Park
NC 27709
USA

M. W. Leete
Flymo Ltd – Electrolux Group
  Manufacturing Co.
Preston Road
Aycliffe Industrial Estate
Newton Aycliffe
Co. Durham DL5 6UB
England

H. Makino
Engineering School
Yamanashi University
4-3-11 Takeda, Kofu-shi
Yamanashi-ken
Japan

J. Miller
School of Engineering
The Hatfield Polytechnic
P.O. Box 109
Hatfield
Hertfordshire AL10 9AB
England

T. Niinomi
PERL Hitachi Ltd
292 Yoshida-Machi
Totsuka-ku
Yokohama 244
Japan

K. Rathmill
Cranfield Robotics and
  Automation Group
College of Manufacturing
Cranfield Institute of Technology
Bedford MK43 OAL
England

A. H. Redford
Department of Aeronautical
  and Mechanical Engineering
University of Salford
Salford
England

P. Saraga
Philips Research Laboratories
Cross Oak Lane
Redhill
Surrey RH1 5HA
England

R. D. Schraft
Fraunhofer Institut für Produktions-
technik und Automatisierung (IPA)
Postfach 800 469
Nobelstrasse 12
D-7000 Stuttgart 80
West Germany

B. E. Shimano
Adept Technology Inc.
Mountain View, CA
USA

J. Simons
Afdeling Mechanische Konstruktie
  en Produktie
Katholieke Universiteit Leuven
Celestijinenlaan 300B
B-3030 Leuven
Belgium

J. Spaa
Department of Applied Physics
Delft University of Technology
P. O. Box 5046
2600 GA Delft
The Netherlands

R. N. Stauffer
Editor, Robotics Today
One SME Drive
P. O. Box 930
Dearborn, MI 48121
USA

K. Sugimoto
PERL Hitachi Ltd
292 Yoshida-Machi
Totsuka-ku
Yokohama 244
Japan

M. Takahashi
PERL Hitachi Ltd
292 Yoshida-Machi
Totsuka-ku
Yokohama 244
Japan

G. J. VanderBrug
Automatix Inc.
1000 Tech Park Drive
Billerica, MA 01821
USA